Introduction to Biotechnology

INTRODUCTION TO BIOTECHNOLOGY

(EBT-104/EBT-204)

AS PER U. P. TECHNICAL UNIVERSITY SYLLABUS

By

NAVEEN DWIVEDI

Assistant Professor
Department of Biotechnology
S.D. College of Engineering and
Technology, Muzaffarnagar
Uttar Pradesh

SHUBHA DWIVEDI

Assistant Professor
Department of Biotechnology
S.D. College of Engineering and
Technology, Muzaffarnagar
Uttar Pradesh

UNIVERSITY SCIENCE PRESS
(An Imprint of Laxmi Publications Pvt. Ltd.)

BANGALORE • CHENNAI • COCHIN • GUWAHATI • HYDERABAD
JALANDHAR • KOLKATA • LUCKNOW • MUMBAI • RANCHI
NEW DELHI • BOSTON, USA

Published by :

UNIVERSITY SCIENCE PRESS

(*An Imprint of Laxmi Publications Pvt. Ltd.*)
113, Golden House, Daryaganj,
New Delhi-110002

Phone : 011-43 53 25 00
Fax : 011-43 53 25 28

www.laxmipublications.com
info@laxmipublications.com

First Edition : 2013

OFFICES

© **Bangalore**	080-26 75 69 30	© **Chennai**	044-24 34 47 26
© **Cochin**	0484-237 70 04, 405 13 03	© **Guwahati**	0361-254 36 69, 251 38 81
© **Hyderabad**	040-24 65 23 33	© **Jalandhar**	0181-222 12 72
© **Kolkata**	033-22 27 43 84	© **Lucknow**	0522-220 99 16
© **Mumbai**	022-24 91 54 15, 24 92 78 69	© **Ranchi**	0651-220 44 64

UIB-9691-150-INTRO TO BIOTECHNOLOGY-DWI C— ???

Typesetted at : Shubham Composer, New Delhi. Printed at : Repro Knowledgecast Limited, Thane

Dedicated to
My adored & respected parents

Contents

PREFACE

Biotechnology is one of the fastest growing, and most arguable area of science and technology in the 21st century. In the last few decades Biotechnology has emerged as a field with great potentials. From the development of Bt-cotton to Human Genome Project, biotechnology has been playing a vital role in the establishment of new milestones in biological research. This book "Introduction to biotechnology" will be useful for undergraduate students of Biotechnology, from different universities/ colleges. This book has been written to meet all academic requirements of the students as per the U.P. Technical University syllabus in a very comprehensive and consistent way. The author does not claim this as his innovative research work. It is just a conclusion of the compilation of the researches done by the scientists in different area of the biotechnology all over the world. During the writing of this book help has been taken from different sources like, review and research papers from reputed journals, some standard text books to revise the knowledge. The subject matter and language of text is very simple and illustrated.

This book comprises five chapters. ***Chapter 1*** deals fundamentals of biochemical engineering, role of biotechnology in society, application of biotechnology, genetic modified organisms and biosafety issues. ***Chapter 2*** describes detail about biomolecules: carbohydrates, proteins, lipids, nucleic acid and enzymes, classification and nomenclature, Enzyme specificity, Enzyme specificity hypothesis, Properties of enzymes, Factors affecting enzyme activity, Purification and Characterization of Enzymes, Industrial applications of enzymes. ***Chapter 3*** deals Introduction about *Prokaryotic* cells and *Eukaryotic* cells, Classification and Applications of microbes. ***Chapter 4*** titled bioinformatics; is the application of information technology to the management of biological data; gives a complete overview on introduction and application of bioinformatics, Biological Data bases (Nucleotide sequence Databases, Protein sequence Databases, Structure Databases), Retrieval of biological data, Sequence file formats and Database searching tools. ***Chapter 5*** describes Gene and genome, History of genomics, Introduction Genome Sequencing Projects, Gene Prediction and Counting, Genome Similarity, SNPs, Types of Genomics. The review questions are given at the end of each chapter.

I anticipate the book will attest a valuable tool for students and teachers. Critical suggestions from users are welcomed for the improvement of the book.

—Authors

ACKNOWLEDGEMENT

First of all we pay our indebtedness to Almighty GOD who gave us idea, strength, and constant patience which is very necessary for doing this work.

I am grateful to Dr. B. N. Mishra (Head, Department of Biotechnology, IET, Lucknow) for introducing me to this field.

We are indebted our profound gratitude to Dr. C. B. Majumder (Associate Professor, Department of Chemical Engineering, IIT Roorkee) and Dr. P. Mondal (Assistant Professor, Department of Chemical Engineering, IIT Roorkee) without whose credence in the worth of this concerted endeavour could not be possible.

We would like to thank Dr. S. N. Chauhan (Director, SDCET, Muzaffarnagar), all authorities of SDCET and all colleagues of the department of biotechnology for their precious advise and support. We are also thankful to Dr. Pradeep Kumar (Assistant Professor, Department of Chemical Engineering, IT-BHU) and our friends for their valuable suggestions and constant support.

We are also grateful to my parents and my son Vedang, who have been inspirational source of strength during all phases of this work.

We are also acknowledge to Laxmi Publication for their support and kind patience in publishing this book.

—Authors

Syllabus

Introduction to Biotechnology
EBT-104/ EBT-204

Unit I: Introduction to Biotechnology

Fundamentals of Biochemical Engineering, Biotechnology and Society. Principles and Processes; Application in Health, food, medicine and Agriculture; genetically modified (GM) organisms; biosafety issues.

Unit II: Biomolecules

Building Blocks of Biomolecules-Structure and dynamics. Structure and function of Macromolecules (Carbohydrates, Proteins, Lipids). Classification of Enzymes; Purification and characterization of enzymes from natural sources. Comparison of chemical and enzyme catalysis.

Unit III: The Cell

Cell as a basic unit of life. Introduction: Definition, Study of Microbes, Types of microbes, Classification of microbes. Origin of microbiology. Application of microbes in fermentation Biotechnology. Cellular Techniques including chromatography.

Unit IV: Bioinformatics

History of Bioinformatics. Introduction and application. Biological databases (nucleotide and protein data bases, Structure databases) and their retrieval. Sequence file formats. Information Sources Analysis using Bioinformatics tools.

Unit V: Genomics

Introduction Genome Sequencing Projects, Gene Prediction and counting, Genome similarity, SNP's and comparative genomics.

1

INTRODUCTION

What is Biotechnology?

The term "**Biotechnology**" refers to the use of living organisms or their products to improve human health and the human environment. Prehistoric biotechnologists did this as they used yeast cells to raise bread dough and to ferment alcoholic beverages, and bacterial cells to make cheeses and yogurts and as they bred their strong, productive animals to make even stronger and more productive offspring. Hungarian engineer, **Karl Ereky** was coined the term biotechnology in 1917, to describe a process for large scale production of pigs. In broad spectrum biotechnology has been defined as "The development and utilization of biological processes, forms and systems for obtaining maximum benefits to man and other forms of life". Some definitions of biotechnology are given below, which helps the readers to know about its nature and scope.

- *Biotechnology is "The science of applied biological process"* ***(Biotechnology: A Dutch Perspective, 1981)***
- *Biotechnology is "The integrated use of biochemistry, microbiology and engineering sciences in order to achieve technological application of the capability of microorganisms, cultured tissues/ cells and parts thereof."* ***(The European Federation of Biotechnology EFB 1981)***
- *Biotechnology consists of "The controlled use of biological agents such as microorganisms or cellular components for beneficial use"* ***(U.S. National Science Federation)***
- *Biotechnology comprises the "Controlled and deliberate application of simple biological agents, living or dead, cells or cell components, in technically useful operation either or productive manufacture or as service operation"* ***(J.D. Bulock, 1987)***
- *Biotechnology is the application of scientific and engineering principles to the processing of materials by biological agents to provide goods and service.* ***(OECD, 1981)***
- *Biotechnology may be defined as "The use of living organisms in systems or process for the manufacture of useful products, it may involve algae, bacteria, fungi, yeast, cells of higher animals and plants or sub systems of any of these or isolated components from living matter".* ***(Gibbs and Green Hhalgh, 1983)***
- *The* ***OECD*** *(Organization for Economic Cooperation and development1999) accepted a working definition of biotechnology, where biotechnologies are the products namely knowledge, goods and services, arising from the alteration of living or non- living materials, through the application of science and technology to living organisms as well as parts, products and models thereof.*

Biotechnology provides a set of tools that, if appropriately integrated with other technologies, can be applied for the sustainable development of agriculture, natural resources, health and pharmaceuticals, industry as well as protection of the environment.

History and Development of Biotechnology

The development of biotechnology is divided into three main generations/phases:

- The first generation biotechnology started in **1750 BC** and included processes such as fermentation, microbial, natural products and the use of biological control agents against animal and plant diseases and pests. The fermentation process is widely used in industries for production of beverages and medicines. The microbial process is used in agriculture for nitrogen fixation, control of pollution in the environment, bio-mining, biogas production and also in pharmaceutical industries.
- The second-generation biotechnology started in **1863**, when Gregor Mendel discovered that pea plants passed on traits from parent to progeny in discrete biological units that would be later known as genes. The second generation also includes tissue culture, polyclonal and monoclonal antibodies, and DNA~markers techniques. Tissue culture is used in agriculture and livestock for rapid multiplication or micro- propagation and production of pathogen free plants, embryo rescue and artificial insemination. Polyclonal and monoclonal antibodies techniques are used in the production of medicines and for vaccine development as well as diagnostic tools for animal and plant diseases. The DNA markers techniques are used agriculture and livestock for characterization of animal and plant genomes, selection processes, and as a diagnostic tool.
- The third generation of biotechnology is known as "Genetic Engineering (GE)", "recombinant DNA technology", "gene technology" or "modern biotechnology" started in **1972** when scientists pioneered a way of combining biochemistry in a technique that led to the birth of recombinant DNA, a modified DNA molecule created by combining DNAs from two unrelated organisms. The technology allows the transfer of selected genes between different organisms, species, genera and phyla. Once transferred, these genes may be transferred to offspring of the modified individual, through normal reproductive processes. Genetic engineering has resulted in the production of Genetically Modified Organisms (GMOs).

History of Biotechnology

1953 Unravelling the double helix: James Watson and Francis Crick deduced the structure of the DNA molecule—a double helix.

1967 Cracking the genetic code: HarGovind Khorana, Robert Holley, and Marshall Nirenberg decipher the mechanism that enables DNA to be translated into proteins.

1970 Restriction enzymes discovered: Boyer isolates the "Big Daddy" of restriction enzymes, EcoR1. In the following years, hundreds of different restriction endonucleases are found that cleave DNA at specific sites.

1972 Recombinant DNA technology begins: Paul Berg splices together two blunt-ended fragments of DNA from the SV40 virus and *E. coli*, creating a recombinant.

1972 Cloning: Cohen and Boyer discuss ways to combine plasmid isolation with DNA splicing the basis of the biotechnology industry.

1975 DNA sequencing developed: Gilbert and Maxam of Harvard University and Sanger of Cambridge University, simultaneously come up with two techniques for determining the exact sequence of bases that make up a gene.

1975 Monoclonal antibody technology developed: Milstein, Kohler and Jeme fused immortal tumor cells with antibody-producing B-lymphocyte cells to produce "hybridomas" that continuously synthesize identical (or "monoclonal") antibodies.

1978 Human insulin cloned into *E. Coli.*

1982 Human insulin or humulin: The first recombinant DNA drug approved by FDA.

1986 The Polymerase Chain Reaction (PCR): Developed by Kary Mullis and revolutionizes molecular biology.

1989 The Human Genome Project (HGP) begins: A plan to "map" human genes by the year 2005. Anticipated cost: $3 billion.

1990 First use of gene therapy to treat human patient: A 4-year-old girl with ADA deficiency is the first recipient of gene therapy.

1994 Brave new foods: FlavrSavr tomato, the first transgenic food introduced to the supermarket shelves.

1996 First mammal cloned from adult cells: Dolly, a lamb cloned from an udder cell of an adult sheep born 6 years earlier.

1996 Development of the gene chip: Breakthrough in gene expression and DNA sequencing technology with the introduction of DNA chips, small glass or silica microchips that contain thousands of individual genes that can be analyzed simultaneously.

2003: The Human Genome Project Completed: The International Human Genome Sequencing Consortium announced the successful completion of the Human Genome Project more than two years ahead of schedule for a cost substantially less than the original estimates.

2003: 50th anniversary of the discovery of the structure of DNA.

Biotechnology: An Interdisciplinary Area:

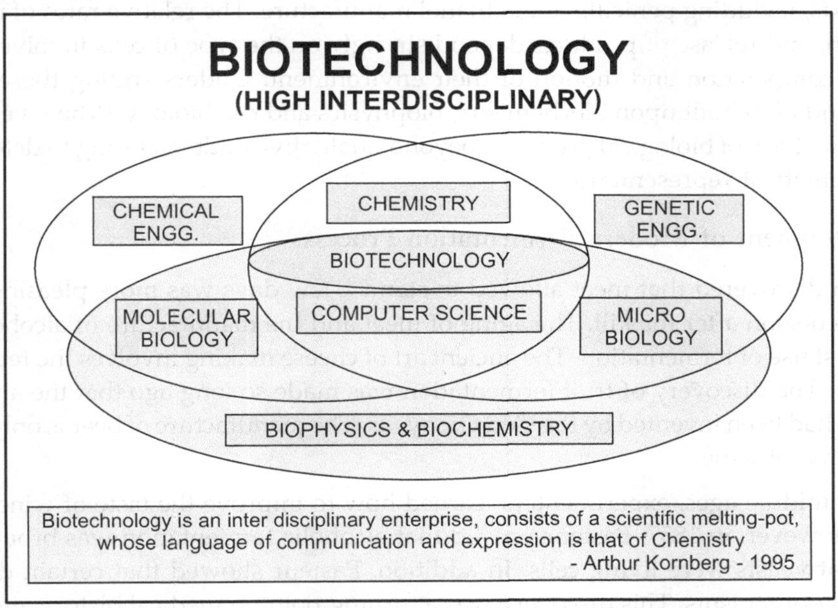

Fig. 1.1 An inter disciplinary area of biotechnology

1.1 FUNDAMENTALS OF BIOCHEMICAL ENGINEERING

Biochemical engineering is concerned with conducting biological processes on an industrial scale, providing the link between biology and chemical engineering. So, we can say Biochemical engineering has usually meant the extension of chemical engineering principle to systems using a biological catalyst to bring about desired chemical transformations.

Biochemical engineers develop bioprocesses which translate biological discoveries into new medicines, groundbreaking stem cell therapies and green sustainable technologies. Biochemical engineering has its origins in traditional chemical engineering and is now the most rapidly developing sector.

There is a difference between bioprocess engineering and biochemical engineering. In addition to chemical engineering, bioprocess engineering would include the work of mechanical, electrical and industrial engineering to apply the principles of their disciplines to processes based on using living cells or subcomponents of such cells. The problems of detailed equipment design, sensor development, control algorithms and manufacturing strategies can utilize principles from these disciplines. Biochemical engineering is more limited in the sense that a draws primarily from chemical engineering principles and broader in the sense that it is not restricted to well defined artificially constructed processes, but can be applied to natural systems.

Our challenge in learning biochemical engineering is to understand and analyse the processes of biotechnology so that we can design and operate them in a rational way. To reach this goal, however, a basic working knowledge of cell growth and function is required. These factors and others peculiar to biological systems usually dominate biochemical process engineering. Finally, the products formed and released during cellular activity are of major concern in many industrial and natural contexts, including penicillin and ethanol manufacture. The relative rates of nutrient utilization, growth, and release of products depend strongly on the type of cells involved and on the temperature, composition and motion of their environment. Understanding these interactions requires a foundation built upon biochemistry, biophysics and cell biology. Whenever possible we shall extent our study of biological processes beyond qualitative understanding to determine quantitative mathematical representation.

1.1.1 Development of Modern Fermentation Process

The cave man discovered that meet allowed to stand a few days was more pleasing to the taste than meat eaten soon after the kill. The aging of meat and the manufacture of alcoholic beverage were man's first use of fermentation. The ancient art of cheese making involves the fermentation of milk or cream. The discovery of fruit fermentation was made so long ago that the ancient Greeks believed wine had been invented by the God Dionysus. The manufacture of beer is only slightly less ancient than that of wine.

During the middle ages, experimenters learned how to improve the taste of wine, bread, beer and cheese. However, in 1857, Pasteur proved that alcoholic fermentation was brought about by yeasts and that yeasts were living cells. In addition, Pasteur showed that certain diseases were caused by microorganisms. This discovery was a turning point in medical history and the birth of microbiology. He concluded the *bacillus* which causes anthrax in cattle could thrive in the tissues of cattle, but appeared to be inhibited by the presence of certain airborne microbes. This prompt Pasteur to suggest that human diseases could be cured by marshalling microbe against microbe. In

1901, Rudolf Emmerich and Oscar Low, of the University of Munich, isolated "pyocyanase" from *Pseudomonas aeruginosa*, a bacterium. Several hundred patients were treated quite successfully with pyocyanase, the world's first antibiotic. In 1923, Pfizer opened the world's first successful plant for citric acid fermentation. The process involved fermentation utilizing the mold *Aspergillus niger* where by ordinary sugar was transformed in to citric acid. Other industrial chemicals produced by fermentation included butanol, acetic acid, oxalic acid, gluconic acid, fumaric acid and many more. In 1928, Alexander Fleming, working with *Staphylococcus aureus*, a bacterium that causes boils. A mold of the *penicillium* family grew as a contaminant on a Petri dish inoculated with *Staphylococcus aureus*: a clear zone was observed, where the *Staphylococcus* organisms in the vicinity of the contaminating mold had been killed. Fleming nurtured the mold and than extracted a chemical from it which killed the bacteria. He named the extracted material penicillin and used most of his meagre supply to clear up one infected wound. From the throat of a chicken professor Sclman A. Waksman of Rutgers University isolated and actinomycete, *Streptomyces griseus*, which elaborated a new antibiotic, streptomycin which is particularly effective against the causative organism of tuberculosis. the list of antibiotics is long today and include such important antibiotics as chloramphenicol, tetracyclines, bacitracin, erythromycin, novobiocin, nystatin, kanamycin etc. Progress in fermentation is continuing at an ever increasing pace. Each year new products are added to the list of compounds derived from fermentation. Several vitamins are now produced routinely employing fermentation steps in their synthesis. Outstanding examples are B-2 (riboflavin), B-12 (cyanocobalamin) and C (ascorbic acid). Fermentative synthesis of the amino acid like L-lysine and L-glutamic acids are also being carried out commercially. Important agricultural uses are being found for the new fermentation product gibberellins, a plant growth regulator; and crystalline inclusions from a species of *bacillus* are being used as specific insecticides in another agricultural application.

1.1.2 Overview of Integrated Bioprocess and its Status in Fermentation Industry

The contribution of engineer for the development of the penicillin fermentation has already been discussed. The outgrowth of the undertaking was the pure culture technique, carried out in aerated and agitated deep tank fermentors. This technique is very similar to its predecessor used for yeast propagation, introduced to the biochemical process industry refined fermentation equipment capable of being maintained under aseptic conditions even when vigorously aerated. Integrated bioprocess consists various steps, some of them sequential and others concurred. All the steps of integrated bioprocess are shown in the Fig. 1.2 and will be discussed here briefly in general.

1.1.2.1 Unit Operations

Fermentation processes have in common many of the familiar chemical engineering unit operations. All the rest of the physical operations carried out on the fermentation broth for product recovery and downstream operations including the effluent treatment may be classified as unit operations. Even, some of the operations carried out on the fermenter such as fluid dynamics, air sparging, mixing, heat transfer to the fluids, etc., may also be categorized as unit operations. Aerobic fermentation involves the "mixing" of three heterogeneous phases: microorganisms, medium and air. Analysis of fermentations by the unit operation technique has added greatly to the understanding of their behaviour. Close cooperation between biologists and engineers is needed now to device logical methods for screening large numbers of strains, and translating the results of shaken-flask and pilot plant experiments to production vessels.

1.1.2.2 *Unit Processes*

A careful analysis required for several industrially reliable fermentation processes shows that there are common reactions for physical and chemical point of view. Fermentation processes can be classified by the reaction mechanism involved in converting the raw materials in to the products; these include reductions, simple and complex oxidations, substrate conversions, transformations, hydrolyses, polymerizations, complex biosynthesis and the formation of cells. More importantly, for the biochemical engineer unit process offers a logical approach to an examination of fermentation reaction mechanisms.

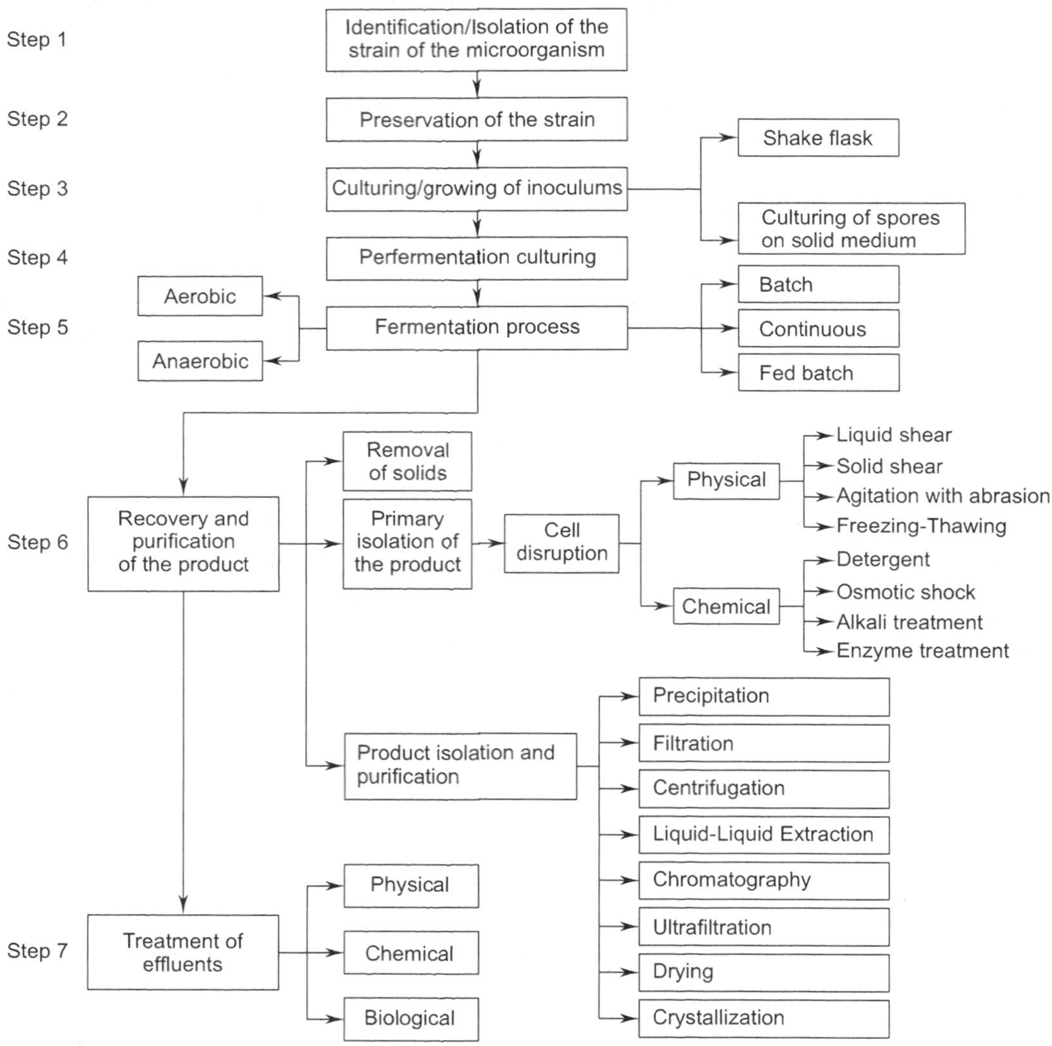

Fig. 1.2 Various integrated bio processing steps

1.1.2.3 *Process Design*

Design for fermentation process requires and understanding of stoichiometry and reaction kinetics. In design for batch fermentation, a consideration of the kinetics of the process often has not been necessary. However, as continuous fermentation has developed, reaction kinetics has become increasingly important. Successful process design depends on successful process control. Although, certain process variables such as temperature, air, flow and agitation were controlled in even the earliest deep tank fermentations, other variables remained unmonitored largely, because of the difficulties of obtaining sensitive measurements by maintaining aseptic conditions. Successful development of control devices for these and other variables will permit greater economy in substrate utilization, power and aeration.

Competition among different fermentation product demands increases in productivity and reductions in the costs of biological processes and the returns from the research needed to develop new processes are being critically assessed. Advanced instrumentation is now being applied to fermentations. In the fermentation industry, the diversity of operations, the ever changing nature of fermentation products, their values, and the complexity of equipment, make costs and especially important consideration; equipments must be versatile as well as efficient and these are difficult aims to achieve simultaneously.

1.2 BIOTECHNOLOGY AND SOCIETY

The effects of any new technology introduced on the scale anticipated for biotechnology extend beyond the industries and research centres influencing our everyday lives. Biotechnology has, for example, made it possible to detect, and in some cases treat, diseases such as sickle-cell anaemia, Tay-Sachs disease, diabetes, and cystic fibrosis. Following initial concerns that genetic engineering could give rise to infectious organisms—the spread of which would be difficult to contain—a stringent set of guidelines was drawn up by the government and leading scientists in the mid-1970s to regulate research in this field. While, it is not possible to eliminate completely the risk of a genetic engineering accident, the experience of the last ten or so years of research have indicated that the chances of constructing a disease-producing organism by accident are very remote. This is because such pathogens require an extremely complex set of distinct characteristics, and are effective only when all are present.

A High-Tech, Low-Risk Technology

Containment of experiments is the key to safety. Microbiologists have gained valuable experience over many years in handling extremely dangerous natural organisms, such as smallpox virus and cholera bacteria. Physical containment (airtight chambers and sterilization of all equipment) is backed up by biological containment. The K-12 strain of E. coli used for the vast majority of experiments, although originating from the E. coli organism present in the human intestine, has become accustomed to laboratory "penthouse" conditions. These optimum conditions are provided by microbiologists concerned with minimizing variation in laboratory data. As a result, scientists have shown that such strains cannot survive in the harsher conditions of the human body or the external environment. Other approaches include the use of strains that specifically require for their survival chemicals not present in the human body. Thus, current research carried out under the strictest guidelines carries minimal risk to workers and the public at large. Concern has been voiced that biotechnology might increase the risk of biological warfare, and some have speculated that biologists today are stepping into the shoes of the nuclear physicists of 40 years ago. It is undoubtedly a daunting aspect of the deployment of biotechnology that will require continued vigilance. Improved genetic tests based on biotechnological advances can be used to track down criminals in assault cases based on the uniqueness of their DNA. Genetic counselling can provide advice on heritable diseases, and genetic screening of workers in possible risk industries is being considered. DNA probes are providing breakthroughs in early diagnosis of disease. As detection of genetic predispositions becomes more predictable, a great deal may be known at birth of an individual's prospects in life. The moral question then arises as to who has access to this information and how this will affect the individual's quality of life.

Is Biotechnology Safe?

Many people have voiced concern about biotechnology and genetic engineering. Scientists have considered the issue of safety over recent years. A special committee of the National Academy of Sciences specifically reviewed the issues on the introduction into the environment of organisms genetically engineered using recombinant DNA technology. They concluded that "there is no evidence that unique hazards exist either in the use of R-DNA technique or in the transfer of genes between unrelated organisms," and that "the risks associated with the introduction of R-DNA engineered organisms are the same kind as those associated with the introduction of unmodified organisms."

Legal Implications

It will be essential that such issues are aired in public debate as the technology develops. Many countries are actively reviewing the safety and ethics of biotechnology research and its applications. Some countries have already established research guidelines for work on embryo transplantation, embryo research, and surrogate motherhood. Lawyers and the public at large will be required to face up to these and similar questions as the biosciences, and biotechnology in particular, move forward.

The Benefits

The potential benefits include solving world food shortages, and improvements in medicine, agriculture, and veterinary sciences. We can confidently expect biotechnological solutions to many essential industrial processes that currently produce toxic effluents. An increasing role for biotechnology in environmental management will undoubtedly follow. Because, the prospect of serious biohazards appears to be receding, it does not mean that strict regulation of the new technology should be relaxed. Provided such vigilance is maintained, mankind can look forward to a wide range of exciting prospects that stem from biotechnology.

1.3 APPLICATIONS OF BIOTECHNOLOGY

There are many essential and innovative products and applications of biotechnology. Biotechnology should be sighted as a spectrum of enabling technologies that establish a diversity of applications in the field of not only products for human use, but we also consider the applications of biotechnology in agricultural, medical, industrial, environmental, microbiology, plant biology and marine biology, among other disciplines. The substantial amounts of biotechnology products currently available are far too abundant, for example pharmaceutical production, creating drugs for the treatment of human health conditions. There are now several hundred drugs, vaccines and diagnostics on the market with more than 300 biotechnology medicines in development targeting over 200 diseases. There are various applications of biotechnology in diverse meadow:

1.3.1 Diseases Diagnosis and Treatment

More accurately and rapidly diagnosis of diseases and medical state of affairs can be done by advancement of biotechnology based tools. Several home testing kits are the best example of the advancement of biotechnology in the area of medical science.

Lectins are proteins, or glycoproteins, that agglutinate erythrocytes of some or all blood groups *invitro*. They are an important group of bioactive proteins and glycoproteins found in most organisms, including plants, vertebrates, invertebrates, bacteria and viruses, and have several important applications to the fields of health food and medical biotechnology. Lectins are used as tools in the fields of biochemistry, cell biology and immunology, as well as for diagnostic and therapeutic purposes in cancer research. Lectins can bind reversibly with free sugars or with sugar residues of polysaccharides, glycoproteins or glycolipids. There has been increasing demand for novel diagnostic and medical cancer therapies that utilize non-traditional sources. Lectins were reported to bind preferentially to cancer cell membranes, or their receptors, causing cytotoxicity, apoptosis and/or tumour growth inhibition.

Biomarker is the diagnostic tool which used to detect most of the diseases at their early stage. For example, cancer, in this disease early detection is done for providing the best treatment: Biomarkers are mainly proteins which are produced by diseased tissue. Many biomarkers are released in to body fluid as a product of cell damage. For *e.g.*, a protein **prostate-specific antigen** is released in to the blood stream when the prostate gland is inflamed. Most of the biotechnology companies are actively involved in searching for better biomarkers that can be used for early detection.

Most of the genetic diseases are diagnosed by using various molecular biology based methods. The purpose of most of the genetic testing on foetus is to identify the sex of a child or to diagnose chromosomal abnormalities. **Chorionic Villus Sampling** (CVS) used for foetal testing. According to this procedure, a suction tube is used to eliminate a small portion of a layer of cells known as chorionic villi. In CVS method, adequate cells are obtained so the sample can immediately use for karyotyping. Advantage of chorionic villus sampling is that the method can be done earlier in the pregnancy, around 8–10 weeks. A new technique evolved for karyotyping in both foetus and adult. This technique is known as **Fluorescence in Situ Hybridisation** (FISH). This technique is very useful for identifying the missing chromosomes and extra chromosomes. In this technique, a chromosome spread is prepared on a slide and then fluorescent probes are hybridized to each chromosome. Each probe is specific for certain marker sequences on each chromosome. For *e.g.*, a disease, chronic myelogenous leukaemia (Fig. 1.3) which is a cancer of white blood cells happened when genes on chromosome 9 and 22 are swapped can be detected by using fluorescence in situ hybridisation techniques.

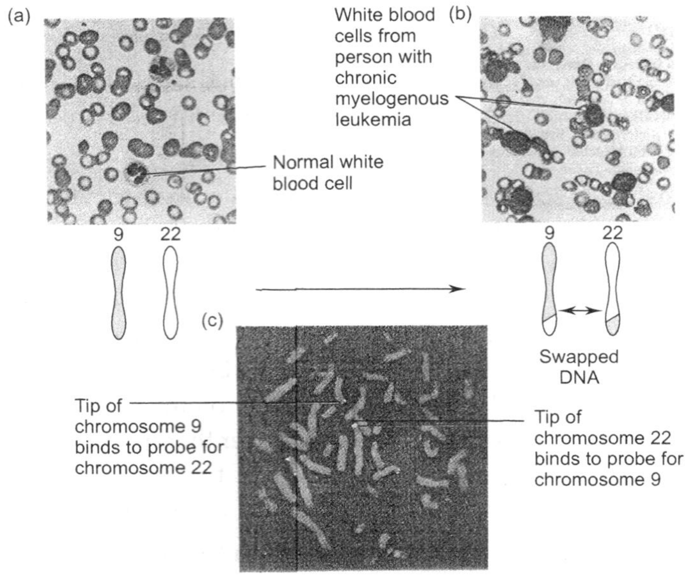

Fig. 1.3 Detection of chromosomal defects by FISH method

Another analysis which has been performed on both that is embryos and adults to find out the genetic disabilities. This technique is known as Restriction Fragment Length Polymorphism (RFLP) analysis. The basic principal of RFLP analysis is that defective gene sequences may be cut differently by restriction enzymes than their normal compliments; the region behind this is nucleotide changes in the mutant genes can affect restriction enzyme cutting sites to create more or fewer. This technique is used for DNA finger printing and also detects sickle-cell disease (Fig. 1.4). Sickle-cell disease happens when a person has two mutant versions of the β-globin gene.

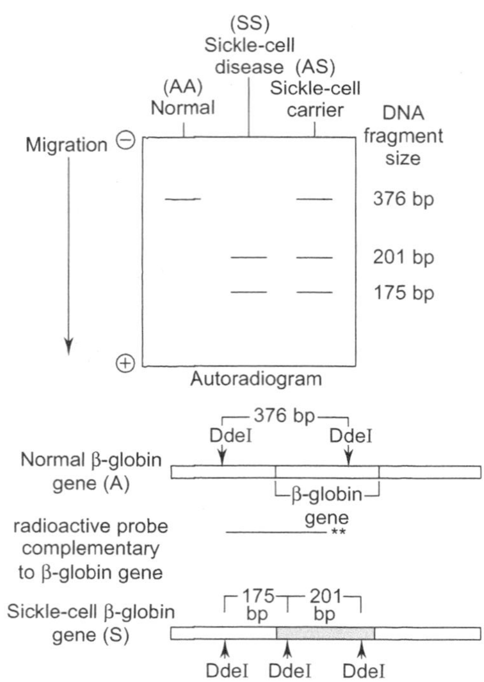

Fig. 1.4 Detection of genetic disease by RFLP method

Gene microarray or **DNA microarray** is a very new technique for studying the genome and it also contribute an important role for detecting genetic disease. Most of the scientist can use microarrays to screen a patient for a pattern of genes that might be expressed in a particular diseases condition. The gene microarray analysis used to find out the patient's risk of developing disease.

Pharmacogenomics

The rapid advance of the Human Genome Project and the development of new genetic analysis technologies promise to bring a new epoch of genomics to medicine (Collins, 1999; Friend, 1999). The applications of biotechnology and genetics to medicine, pharmacogenomics will likely be one of the first substantial benefits from the Human Genome Project. It has been suggested that the use of pharmacogenomics will be widespread and lead to overall decreased health care costs. Pharmacogenomics is the use of genomic technology (high capacity molecular genetics) to study the genetic differences in how people respond to medical treatment. Many medical treatments work better in some people than others; some do not work at all in some patients. Some of this difference is caused by genetic differences. For *e.g.*, patient's response to a chemotherapy compound called 6-mercaptopurine, which has been used to treat persons with a form of blood cancer called acute lymphocytic anemia. This shows that different persons with the same disease generally respond differently to a drug 6-mercaptopurine, because of slight differences in gene expression. The dose which is given to one person may be toxic for another. By using pharmacogenomics, it has been decided that a proper dose of 6-mercaptopurine is given to the suffered persons (Fig. 1.5). Pharmacogenomics seeks to find the relevant genes and use that knowledge to find which people will respond to a medicine. The use of pharmacogenomics to individualize drug therapy will offer clinical and economic benefits. However, these benefits must be weighed against the additional cost of genotyping all patients in order to adjust therapy in a few. The field of pharmacogenomics is still in its infancy. Its use is currently quite limited, but new approaches are under study in clinical trials. In the future, pharmacogenomics will allow the advance of modified drugs to treat a wide range of health problems, including *Alzheimer disease*, cardiovascular disease, asthma, HIV/AIDS, and cancer.

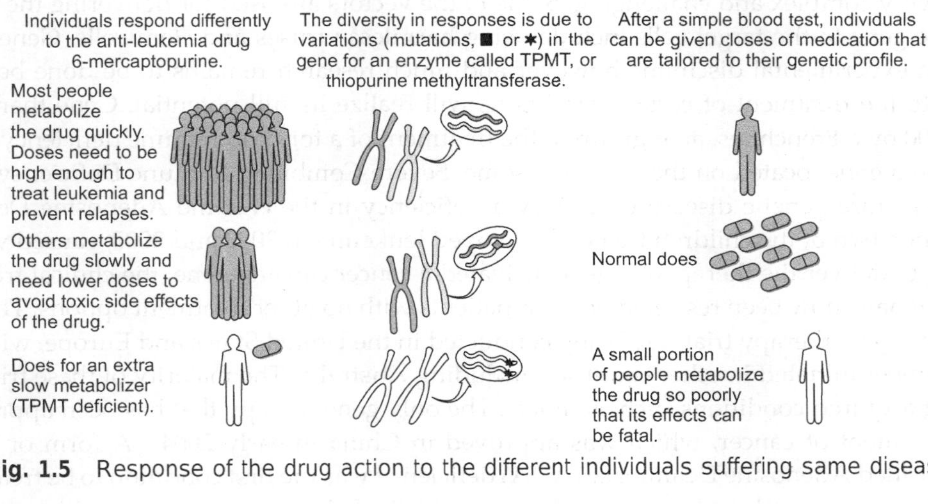

Fig. 1.5 Response of the drug action to the different individuals suffering same disease

Gene Therapy

Gene therapy is altering the genetic framework of a human. The use of genes as a remedial form is called Gene therapy or another word, Gene therapy is a novel pharmacological approach in which the drug is supplied in the form of a nucleic acid DNA, RNA, or some modification or combination thereof. The transfer of a therapeutic or working gene copy into specific cells of an individual in order to repair a defective gene copy (Fig. 1.6). Thus, it may be used to replace a faulty gene, or to introduce a new gene whose function is to cure or to positively modify the clinical course of a condition. There are two approaches, germ-line gene therapy and somatic cell gene therapy. The former changes the "germ cells", the cells which make sperm or ova. This has a permanent effect on all the individuals which are the offspring of whoever had a therapy. Somatic therapy alters the somatic cells, all the non germ cells in the body. Changing them does not affect the germ cells, but does affect the engineered person.

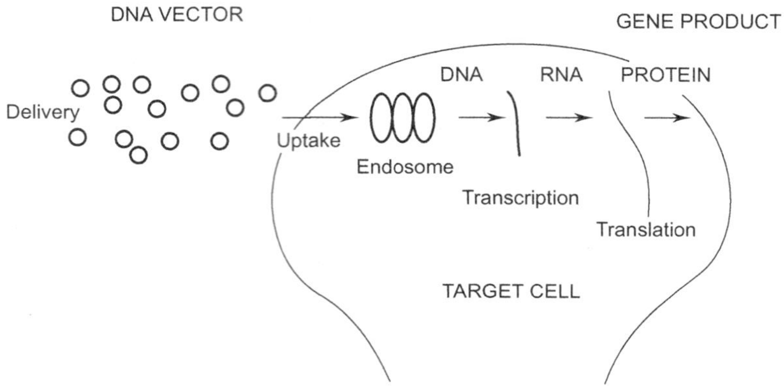

Fig. 1.6 Different steps in gene transfer and expression

The scope of this new approach to the treatment of a condition is broad, with potential in the treatment of many genetic conditions, some forms of cancer and certain viral infections such as AIDS. The problem of 'gene delivery' *i.e.*, how to get the new or replacement genes into the desired tissues, is very complex and challenging. Some of the vectors are used for delivering the working copy of the gene to the target cells include using harmless viruses and stem cells. Gene therapy remains an experimental discipline however and much research remains to be done before this approach to the treatment of certain conditions will realize its full potential. Gene therapy was used in 2000 by a French research group in the treatment of a form of immune deficiency due to a mutation in a gene located on the X chromosome, Severe Combined Immune Deficiency, (SCID), SCID, a very rare genetic disease caused by a deficiency in the enzyme Adenosine Deaminase (ADA). When two of the children treated developed leukemia in 2002 and 2003, caused when the virus used to deliver the therapeutic gene activated a cancer causing gene, the clinical trials were stopped but have now been resumed only for patients with no other treatment options. The majority of clinical gene therapy trials are being conducted in the United States and Europe, with only a modest number initiated in other countries, including Australia. The majority of these trials focus on treating acquired conditions such as cancer. The only gene therapy that has been approved for routine treatment of cancer, which was approved in China in early 2004. A form of immune deficiency called Adenosine Deaminase (ADA) deficiency was the first condition to be treated with a gene therapy approach in humans in the early 1990s. It is also the first condition for which therapeutic gene transfer into stem cells has been attempted in the clinical field.

The completion of the Human Genome Project in 2004, along with the sequencing of nonhuman genomes, has spawned an incredible revolution in the understanding of genetics. Simultaneously, geneticists have raced to develop medicines to treat and cure diseases caused by genes gone awry. Gene therapy treatment that gets at the direct cause of genetic disorders is sometimes touted as the magic bullet, the cure all for inherited diseases like **Cystic fibrosis, Sickle cell anemia, Tay-Sachs, Hemophilia** and **cancer**. Gene therapy may even provide a way to block the genes of pathogens such as the virus that causes AIDS, providing reliable treatments for illnesses that currently have none.

Common Innate Diseases

Mutation is a common occurrence; most inherited diseases are comfortingly rare. Inherited disorders are often recessive and show up only when an individual is homozygous for the trait. Inherited diseases aren't nonexistent, nevertheless. The following relatively common inherited diseases are described.

Cystic fibrosis

The most common inherited disorder among Caucasians (white or light skinned race) in the United States is Cystic Fibrosis (CF); this autosomal recessive disorder occurs in roughly one in every 3,000 births (autosomal recessive means the gene is not on a sex chromosome and a person must have two copies of the allele to get the disease). The mutations that cause CF occur in a gene located on chromosome 7. Persons affected with CF produce thick, sticky mucus in their lungs, intestines, and pancreas. The gene implicated in CF, called the *cystic fibrosis transmembrane conductance regulator gene* (or CFTR for short), normally controls the passage of salt across cell membranes. Water naturally moves to areas, where salt is more concentrated, so the movement of salt from one place to another has an effect on how much water is present in parts of the body. In persons with CF, the removal of salt from the body (via sweat) is abnormally high. As a result, the lungs, pancreas, and digestive system can't retain enough water to dilute the mucus normally found in those systems, so the mucus produced by persons affect by CF is unusually thick. The buildup of thick mucus blocks breathing passages and makes waste elimination difficult, causing severe breathing and digestive difficulties and a high susceptibility to respiratory illnesses. CF is diagnosed in two ways:

- Persons who may be carriers for the mutated allele can be tested genetically
- Children possibly affected by the disease are diagnosed by a "sweat test." Their sweat is tested for salt content, and abnormally high amounts of salt indicate that the child has the disease.

CF is a target of gene therapy, but it resists a cure. Most afflicted persons must endure a lifetime of therapy that includes having someone strike on their chests so that they can remove the mucus from their lungs by coughing

Sickle cell anemia

Sickle cell anemia is the most common genetic disorder. Peoples are generally affected by this autosomal recessive disorder. The mutation responsible for sickle cell is found in chromosome 11, the gene responsible for making one part of the protein complex that composes hemoglobin. In the case of sickle cell, one base is mutated from adenine to thymine (a *transversion*). The mistake changes one amino acid added during translation from glutamic acid to valine, producing a protein that folds improperly and can't carry oxygen effectively. The red blood cells of persons affected by sickle cell take on the disease's characteristic crescent shape when oxygen levels in the body are lower than usual (often as the result of aerobic exercise). The sickling event has the side effect of causing blood clots to form in the smaller blood vessels (capillaries) throughout the body. Clot formation is extremely painful and also causes damage to tissues that are sensitive to oxygen deprivation. Persons with sickle cell are vulnerable to kidney failure, yet with good medical care, most affected persons are expected to live into middle adulthood (40–50 years of age).

Tay-Sachs

An autosomal recessive disorder, Tay-Sachs is a progressive, fatal disease of the nervous system and is unusually common among persons of Ashkenazi (Eastern European) Jewish ancestry. One in every 30 to 40 persons of Jewish ancestry is a carrier of Tay-Sachs. French Canadians and persons of Cajun (south Louisiana) descent are also often carriers of the mutated allele. The mutation that causes Tay-Sachs is found in the gene that codes for the enzyme hexosaminidase A (HEXA). Normally, your body breaks down a class of fats called gangliosides. When HEXA is mutated, the normal metabolism of gangliosides stops, and the fats build up in the brain, causing damage. Children inheriting two copies of the affected allele are normal at birth, but as the fats build up in their brains over time, these children become blind, deaf, mentally impaired and ultimately paralyzed. Most children with Tay-Sachs do not survive beyond the age of 4. Unlike some metabolic disorders, such as phenylketonuria, changes in diet do not prevent the buildup of the unwanted chemical in the body.

Stem Cell Engineering

Presently, stem cell is a thrust area of medical field in order to treatment of human degenerative diseases. At present, the science underlying stem cell-based technologies has a high public profile with intense prospect for success in the future.

Stem cells are undifferentiated cells that have the capacity to self renew and to achieve multilineage differentiation. Stem cells have specific enzyme and cell specific antigens and have the ability to express developmentally regulated gens. Stem cells have been characterized as totipotent, pluripotent and multipotent. In order to know about stem cell it is essential to understand the cellular details of early human development from the fertilized egg to an embryo and then a foetus. The zygote divides rapidly and after 3–5 days it forms a compact ball of about 12 cells to 16 cells called a morula. At this stage, all of the cells retain the potential to form an embryo. These cells are totipotent by nature. In 5 to 7 days, a sphere of specialized cells has been developed which called as blastocyst. The outer ring of the cells of the blastocyst (Fig. 1.7) later will form the placenta and other tissues while the inner cell mass or epiblast will go on to form all of the cells and tissues of the embryo and the human body. This inner cell mass is the source of human embryonic stem cells and these cells are pluripotent in nature.

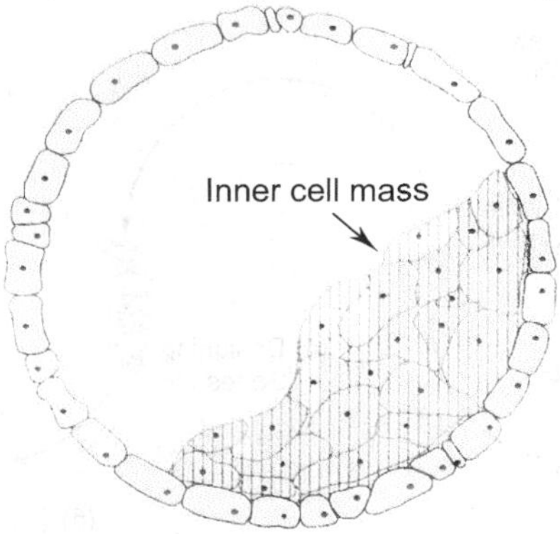

Fig. 1.7 The blastocyst

1.3.2 Agriculture

Agriculture is the largest contributor to the economies of many countries, generating more than half of the annual Gross Domestic Product (GDP) for several countries. The livelihoods of most rural and low-income communities in the countries are to a large extent based on agriculture. Agricultural biotechnology, which comprises a wide range of biological disciplines, offers enormous potential to speed up the development of plant varieties with pro-poor traits such as drought tolerance, pest resistance or tolerance, higher yields, increased nutritional value, among others. Similarly in animal production there is substantial opportunity for development of vaccines and diagnostics targeting diseases which constrain livestock production in developing regions of the world. In addition, genetic markers can aid breeding of livestock for important traits such as disease resistance, improved product quality as well as improved productivity.

Agricultural biotechnology is one of the newest applications of biotechnology, with industries interested in employees with skills in biotechnology and agriculture. Remarkable achievement has been achieved in the years, since World War II to the development of higher yield crops. Particularly for wheat and rice, the improvements in yield have been dramatic and have helped feed an ever growing world population. The first wave of such high yield crop came about through traditional breeding practices using Mendelian genetics to identify and select strains of crops that exhibit particularly valuable traits. Use of Agricultural biotechnology is not only to create crops with extremely valuable traits, but also to develop tools and techniques of great benefit. One of the most beneficial techniques shown in Fig. 1.8 is tissue culture technique. Through the controlled use of plant hormone, it is possible to regenerate whole plants from individual cells that are maintained in culture. *Agrobacterium tumefaciens* has been used particularly to get DNA in to trees by DNA transfer mechanism. Trees are difficult plants to breed, because of their size and long life cycle, and so genetic engineering techniques offer unusual advantages of speed and the ability to engineer millions of clone. Walnut, popular, apple and plum trees have all had DNA transferred to them using *Agrobacterium tumefaciens.*

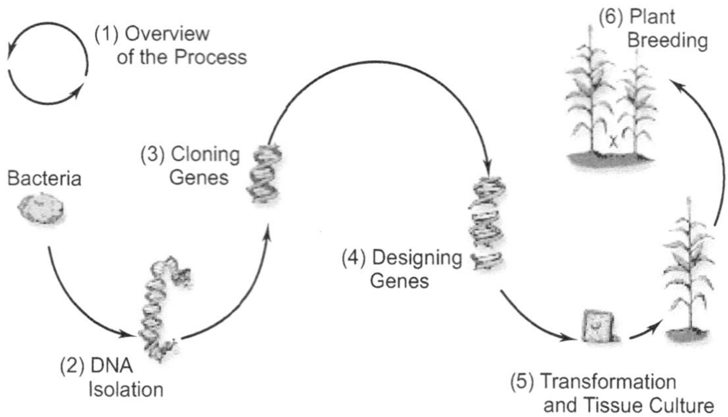

Fig. 1.8 An overview of the tissue culture process

In agriculture, advances in biotechnology have resulted in improved research leading to: drought resistant crop varieties; increased pest and disease resistance in crops and livestock; new, refined diagnostics and vaccines for livestock diseases (*e.g.*, Foot-and-Mouth disease and East Coast fever); rapid propagation of clean planting material (*e.g.*, flowers, vegetables, bananas etc.). Current constraints to the research and application of biotechnology in developing countries include: lack of policies; lack of human and financial resources; lack of public and private investments at levels that can make a difference; absence of systems for the delivery of technologies to potential users; lack of awareness, leading to misconceptions about the potential of, and risks posed by, biotechnology.

Position of Agricultural Biotechnology in Developing Countries

There is now ample evidence to demonstrate the opportunities offered by biotechnology in developing economies. Two examples are given here, from India and China.

India

India's National Dairy Development Board (NDDB) oversees improved dairy production by millions of smallholder livestock producers. Success in using increased milk production to generate increased income on a daily basis is the result of investments in S and T targeting: improved feeding and nutrition of dairy cattle; vaccines to control endemic diseases; and improved animal genotypes and their delivery to farmers. The NDDB organizes delivery of services (including biotechnologies) at the points of milk collection. Payments for technical services are affordable and deducted from milk payments to smallholders. In the 1960s and 1970s India regularly had famines and was a net food importing country. The Green-Revolution in crops and the White Revolution in dairy production are the result of investments in S and T and infrastructure, especially irrigation and communications and the formulation and implementation of supportive public policies (prices, trade etc.) to encourage farmers to go into production. India is now using its productive agricultural sector to guarantee food security and is moving towards industrialisation. Although, there are still millions of people living in poverty in South Asia the trends in India and some of its neighbours are heading in the right direction, with millions moving out of poverty each year.

China

In the early 1980s Chinese leaders decided that science and technology (S and T), especially biotechnology, would be one of the drivers to improve the agricultural sector, and committed substantial public investments in biotechnology, *e.g.*, rice biotechnology (mapping rice genome) and rice breeding (to develop hybrid rice varieties), cotton biotechnology for insect resistance, production of value-added horticultural crops, and complimentary innovations such as use of nematodes for biological pest control leading to increased export markets. Currently, hybrid rice accounts for over 30% of rice in China and over 5 million small farmers are growing Bt-cotton on 1 million hectares of land. Use of biological control has reduced pesticide use on cotton by 30% nationally. Today, horticultural exports are expanding. Thus, through a deliberate effort to revolutionize agriculture, China is making quantum leaps in agricultural productivity and sustainability improvements. The country is now moving to the 'post-Green Revolution era' towards becoming an industrialised nation. It has been pointed out above, an efficient agricultural sector ensures food security and enables industrial development.

1.3.3 Foods

Biotechnology can be used more routinely to monitor, control and enhance quality in all food products, and to understand how food can be kept fresh and palatable for the consumer. The challenge is to recognise the potential of biotechnological techniques to fulfil the food requirements of today's society both for developed and developing countries. Food biotechnology encompasses a wide range of options for improved quality, nutrition, safety and preservation of foods. Modern biotechnological techniques will have considerable importance in influencing trends in the food market, *viz.* Taste, preservation, consistency, colour, cost, safety and health aspects.

A wide range of chemical components of food, such as vitamins, colouring agents, modified starches, fats and lipids, High Fructose Corn Syrup (HFCS) and dairy products like milk, cheese, butter etc are produced by the biotechnological approach. The most extreme example of this is Single Cell Protein (SCP), producing bulk protein for food use using fermentation.

Single Cell Protein

The term single cell protein refers to protein biomass used as a food additive for humans and animals. The main goal of SCP production is to maintain protein quality and quantity. New agricultural practices are widespread, high protein cereals have been developed, the cultivation of soybeans and groundnuts is ever-expanding, protein may be extracted from liquid wastes by ultrafilteration and now the use of microbes as protein producers has gained wide experimental success. The field of study has become known as single cell protein or SCP production, referring to the fact that most of the microorganisms used as producers grow as single or filamentous individuals rather than as complex multicellular organisms such as plants or animals. With humans it has been considered as a protein supplement, as a food additive to improve flavour, fat binding, and more recently as a replacement for animal protein in the diet, because humans have a limited capacity to degrade nucleic acid. Additional processing is required before single cell protein can be used in human food. The advancement of biotechnology contributes to development of single cell protein in large scale worldwide. Research and development in to SCP process has involved work in the fields of food technology, animal nutrition, chemical and process engineering, microbiology, biochemistry, genetics, ecology, toxicology, medicine and economics.

Probiotics

Probiotics is the utilization of living microbes as a health adding part of a food. The living microorganisms have a beneficial consequence on human and animal health by influencing qualitatively and quantitatively the composition and metabolic activity of the microbial flora of the gastro intestinal tract. The average human body contains ten times as many bacterial cells as human cells and they exist in competition with each other, mostly in our guts and on our skin. Disease is often caused by too many of a harmful bacterium, and so Probiotics seeks to promote the growth of other bacteria to out compete the bad once. Probiotics can be produced as capsules, tablets, powders or liquid suspensions. They are enormously used in certain animal/ poultry feeds with definite productivity yield gains. Recently, the sales of bioyoghurts has been increase rapidly that use specific live intestinal bacteria, mostly strain of *Lactobacillus* and *Bifidobacterium*, as starter cultures. Probiotics dairy foods are well established, and most likely do have health conferring value in keeping bad bacteria from our intestine. They can also actually support normal gut function. Some dietary supplements, like fructooligosaccharides, inulin and lactulose are compounds that have beneficial stimulatory effects on probiotic microbial growth and are termed prebiotics.

1.3.4 Bioremediation/Biodegradation

Bioremediation is the degradation or transformation of contaminants into non-hazardous or less hazardous chemicals using microorganisms. Bacterias are normally used for bioremediation, but fungi, algae and plants have also been used for bioremediation. Bioremediation is not a new technology. There has been evidence that compost heaps existed as extreme back as 6000 BC, and in 1891 the first biological sewage treatment plant was created in Sussex, UK. However, the word "bioremediation" did not emerge in peer-reviewed scientific literature until 1987. The bioremediation is classified by three ways:

Biotransformation : The alteration of contaminant molecules into less or nonhazardous molecules.

Biodegradation : The breakdown of organic substances in smaller organic or inorganic molecules.

Mineralization : is the complete biodegradation of organic materials into inorganic constituents such as CO_2 or H_2O.

These three classifications of bioremediation can occur either *in situ* (at the site of contamination) or *ex situ* (contaminant taken out of the site of contamination and treated elsewhere). There are advantages and disadvantages to both in situ and ex situ strategies. Biological treatment method may be aerobic or anaerobic. Biological treatment methods have many advantages over other conventional treatment processes. Some of them have been discussed below:

- Biological treatment process is less energy intensive; it works at mild temperature and ordinary pressure conditions.
- Biological treatment removes a wide variety of pollutants in a single step while Physico-chemical treatment can remove specifically a single or small group of pollutants.
- Risks of explosion, corrosion, toxic gas leakages, etc. are minimal in biological treatments.
- The capital investment is low in biological treatment, due to less complex equipment is required to maintain the desired conditions.

Primarily, bioremediation depends on the use of microorganisms to degrade hazardous waste. This process involves biochemical reactions or pathways in microorganism that result in activity, growth, and reproduction of that organism. The microorganisms also degrade molecules as a source of simple compounds of an element, such as carbon, to synthesize cellular compounds needed for life. Various nonphotosynthetic organisms obtain their energy from organic molecules by oxidation reduction reactions. The organic molecules serve as a source of energy through their oxidation finally to carbon dioxide. Many microorganisms use molecular oxygen as the electron acceptor (oxidant) for the oxidation of organic molecules referred to as aerobic metabolism. In general, the oxidative metabolism involves many sequential reactions referred to as a pathway which results in the addition of oxygen atoms and/or the loss of hydrogen atoms from the organic molecule. In the process, oxygen is reduced and the organic molecule is oxidized as represented by the following reaction 1.

$$(CH_2O)_n + {}_nO_2 \longrightarrow {}_nCO_2 + {}_nH_2O \quad (1)$$

Under anaerobic conditions, many microorganisms get their energy from the oxidation of inorganic and organic compounds. In anaerobic metabolism, inorganic chemicals such as nitrate, NO_3^-, sulfate, SO_4^{2-}, and metals such as Fe^{3+} and Mn^{4+} serve as the electron acceptors for the oxidation. In these processes N_2, H_2S and reduced forms of the metals Fe^{2+} and Mn^{2+} are products of the metabolism as represented by the following reactions 2 and 3.

$$2(CH_2O) + SO_4^{2-} + 2H^+ \longrightarrow 2CO_2 + H_2S + 2H_2O \quad (2)$$

$$5(CH_2O) + 4\,NO^{3-} + 4\,H^+ \longrightarrow 5\,CO_2 + 2\,N_2 + 7\,H_2O \quad (3)$$

In the absence of oxygen, some microorganisms carry out another type of metabolism, in this metabolism there is no external electron acceptors are required. The organic molecule serves both as the electron acceptor and electron donor. The products of the metabolism correspond to both more oxidized and more reduced forms of the organic molecule such as carbon dioxide and methane.

1.4 GENETICALLY MODIFIED (GM) ORGANISMS

Modern Biotechnology offers great hope for directly addressing problems of the developing countries in human health, agriculture and environment. Genetic engineering methods improve the nutritional quality of crops by transferring new genes into indigenous crop strains and confer tolerance or resistance to biotic and abiotic stresses. Similarly, genetic techniques can be used for development of vaccines and drugs against major viruses and parasites infecting humans and livestock. The potential of biotechnology to increase food yields and alleviate hunger and malnutrition is well known. However, little is known about its negative potential for releasing harmful genetically modified organisms in the environment, and for its propensity to displace traditional varieties.

Genetically Modified Organisms (GMOs) can be defined as organism in which the genetic material has been altered from the way that it occurs in nature.

Food ingredients, processing aids, dietary supplements and veterinary chemicals derived from GM microorganisms

Many enzymes used as processing aids in food and feed production are derived through the use of GMMs (European Commission 2004). This means that the GM microorganisms are inactive, degraded or removed from the final product. GM yeasts, fungi and bacteria have been in commercial use for this purpose for over a decade. Examples include: alpha-amylase for bread-making, glucose isomerase for fructose production, and chymosin for cheese-making. Most of the microorganisms modified for food processing are derivatives of microorganisms used in conventional food biotechnology.

GMMs are also permitted in a number of countries for the production of micronutrients, such as vitamins and amino acids used for food or dietary supplement purposes. An example is the production of carotenoids (used as food additives, colourants or dietary supplements) in GM bacterial systems. In the future, complete metabolic pathways could be integrated in GM microorganisms, enabling them to produce new compounds. For animal husbandry, veterinary products such as bovine somatropin, used for increasing milk production, have been developed using genetic engineering. Bovine somatropin has been on the market in several countries for over a decade. The technique of protein engineering aims at altering the genetic, and thus amino acid, sequence of enzymes. Hitherto, protein engineering has not been used extensively in enzyme production. R&D in this area aims to change enzyme characteristics, *e.g.*, improve temperature or pH stability. Enzymatic processing often replaces existing chemical reactions. In many instances, this results in lowered energy consumption and less chemical waste.

Emerging Trends in Genetic Engineering (GE) Research

The emerging scientific and technological developments are enabling complex traits to be addressed, with the intention of developing new products of potential value for agriculture, human health and the environment. Genetic engineering or transgenic research is now focused to complex traits as categorized in Table 1.1. These include:

- Increasing sustainable agricultural production by the cultivation of crops that are better adapted to biotic stresses (pests, diseases and weeds) and abiotic stresses (drought, salinity, and temperature stress).
- Increasing health benefits through more nutritionally beneficial foods, with higher content of essential vitamins and minerals, especially in staple crops such as rice. Reducing allergenic, carcinogenic and/or toxic compounds in certain plants may also be possible, so that they are safer sources of food (*e.g.*, reduced cyanide content in cassava; removing allergenic content of nuts; modifying oil content of certain plants to produce more long chain, poly—unsaturated fatty acids).
- Using plants for pharmaceutical production: certain plants can be developed to produce specific proteins, vaccines against human and animal diseases and other pharmaceuticals.
- Using plants for industrial purposes: production of biodegradable plastics, starch and alcohol production etc.
- Using plants and microbes to mitigate the effects of industrial pollution (bioremediation) by increasing their ability to remove and/or breakdown toxic compounds in the soil (Table 1.1).

Table 1.1 GM crops developed to address complex traits

Category/target	Trait	Transgenic crop
Improved productivity	Drought tolerance Salinity tolerance Water logging Aluminium Tolerance Disease, resistance	Maize Rice Rice Tobacco Rice
Health and nutrition	Vitamin A content Iron content Reduced toxins	Rice, mustard Rice Cassava
Value added traits	Colour changes Flavour changes Shelf life	Flowers Tomato Tomato
Plants for medicinal purposes	Vaccine production	Banana, Potato, tomato, Tobacco
Plants for industrial purposes	Biodegradable plastic production Starch production Alcohol production Starch	Maize Maize Sugarcane Cassava
Plants for Biofuel	Alcohol production	Maize Sugarcane Cassava
Self regulating plants	Limiting gene flow to related and/or wild species	Oilseed rape
Removing toxic compounds from environment (bioremediation)	Mercury pollution Cadmium contamination	Arabidopsis thaliana Tobacco

1.5 BIOSAFETY ISSUES

Biosafety is used to describe efforts to reduce and eliminate the potential risks resulting from biotechnology and its products. It has similarly been defined as "*the avoidance of risk to human health and safety, and to the conservation of the environment, as a result of the use for research and commerce of infectious or genetically modified organisms*" (Zaid, 2001). Relevant scientific disciplines that underpin biosafety studies include molecular biology, plant breeding, genetics, plant pathology, agronomy, weed science, entomology and ecology, among others. Therefore, a large amount of scientific knowledge and data has a direct impact on biosafety, and it can be a difficult process to collate this information in a credible way in order to create a balanced view.

The development of efficient and effective biosafety systems is important not only to accelerate the growth of science and technology and in particular the application of R&D in biotechnology in the region, but also to ensure safe access to new products and technologies developed elsewhere. The absence of a suitable regulatory framework hinders the ability of both the public and the private sectors to invest in new technologies within a particular country and to make new products available to the society. The ICGEB (International Centre for Genetic Engineering and Biotechnology) has also played an important role in issue related to biosafety and the environmentally sustainable use of biotechnology. Since September 1998, the ICGEB has provided an on-line bibliographic data base on biosafety and risk assessment for the environmental release of GMOs. The data base which is accessible through the website of ICGEB also provides informations on biosafety to its member states. The ICGEB is also assisting to its member states in developing the national biosafety framework. Since February 1999, it has also adopted a legally binding biosafety protocol by the signatory countries (ICGEB, Activity report, 1998).

Biosafety: International Efforts

Heightened global environmental awareness and concern over accelerating ecological degradation during the latter quarter of the 20th Century resulted in a desire by the international community to push the protection of the environment higher up the political agenda. These efforts came to completion in 1992 when the **Convention on Biological Diversity (CBD)** came into force. Its objectives include "the conservation of biological diversity, the sustainable use of its components and the fair and equitable sharing of the benefits arising out of the utilisation of genetic resources". During the elaboration of the Convention, negotiators recognised that biotechnology could contribute to achieving these objectives, if developed and used with adequate safety measures for both the environment and human health. Accordingly, procedures were developed to address the safe transfer, handling and use of any LMO ("Living Modified Organism"; used interchangeable with "Genetically Modified Organism" [GMO]) resulting from biotechnology that may have an adverse effect on the conservation and sustainable use of biological diversity (**Article 19.3, CBD**). These procedures formed the **Cartagena Protocol on Biosafety**, which came into force in 2003 and has 138 signatory countries to date (January 2007). Parties lacking a cohesive biosafety policy undertook, or are currently undertaking, a number of initiatives to put a national framework in place in order to comply with the CPB. This has resulted in a great demand for biosafety-related information, training programmes, and capacity-building projects.

Biosafety Concerns and Issues Associated with GMOs

As pointed earlier, the advent of modern biotechnology offers tremendous potential benefits to developing countries, at the same time, its introduction carries with it potential risks. Products of traditional biotechnology, *i.e.*, Fermentation, tissue culture and molecular breeding have been in use in the region for many decades. It is the development and introduction of GMOs that raises a number of legitimate food and health safety, environmental, socio-economic, political and ethical concerns. These concerns have formed the basis for the development and implementation of biosafety frameworks aimed to ensure the safe application of biotechnology.

Food, Animal Feed and Health Safety Issues

Genetic modification has the potential to confer health benefits through, for example:

- Increasing the nutritional content in some crops (*e.g.*, protein-potato being developed in India and vitamin A 'golden rice' in Switzerland.)
- Decreasing the levels of natural toxic compounds in some foodstuffs.
- Reducing the use of pesticides and herbicides in agriculture, with corresponding reductions of their residues in crops.

On the other hand, there is concern that genetic modification could affect the safety of food and animal feed and thus pose potential risks to human and animal health. The GM food introduced proteins and other resulting molecules may cause allergic reactions or act as toxins or carcinogens. In addition, inserting new genes may change the nutrition of crops or their digestibility. Allergenicity could increase either by raising the levels of naturally occurring allergens or through the introduction of new allergens. The great majority of *natural allergens* that affect human beings occur in certain food groups, which have been part of our food system for a long time (*e.g.*, Soya bean, plant nuts, tree nuts, wheat, eggs, fish and shellfish). In the development of GM food crops, close attention is being paid to the issue of allergenicity.

Pesticide toxins introduced in to plants through genetic engineering (*e.g.*, Bt-protein) generate proteins, which may be safe for other animals but carry unexpected **allergenic potential** for humans. 'StarLink' maize is a case in point: the heat tolerant Bt-protein introduced into this GM crop was thought to have allergenic potential in humans and was there initially approved for animal consumption only, by the United States Food and Drugs Administration (USFDA). Conversely, while a GM crop may be safe for humans; its residue used as feed may pose a risk to animals. For example, Bt-cotton is grown commercially in China, India and Indonesia and has passed thorough safety testing. Even so, it is uncertain what long-term impact the widespread use of GM cotton seeds as feed could have on the health of livestock.

The techniques for testing the presence of allergens, toxins and carcinogens in food and feed are well established and are used to test all GM food crops before they are approved for use. Developing countries need to acquire and use these techniques an integral part of any effort they may make in the field of GM crops. They should also to the extent possible use available data, since it is food and feed safety issues of GMOs are universal. This means that GM food and GM feed found to be safe for human and domestic animal consumption in one country are likely to be equally safe for consumption in other parts of the world. Another area of concern is **antibiotic resistance**. The technique of using antibiotic resistant genes as selectable markers in GM plants could result in these genes being transferred to microorganisms that are human pathogens, rendering them antibiotic-resistant. Recognition of this risk is resulting in the phasing out of antibiotic markers or their replacement by others that can be removed from the plants before commercial approval is given.

Environmental Safety Issues

The potential reductions in the use of some chemical pesticides and herbicides that the cultivation of GM crops may allow would have a positive impact on the environment, as would the ability to grow more food on less land. Nevertheless, concern persists about the potential risks posed by GM agriculture to ecology and environment. Harm could potentially arise, directly or indirectly, through six routes:

- Gene flow and transfer of genetically engineered traits to other species
- Invasiveness, weediness and resistance
- Impact on subsoil and other 'non-targeted' organisms, including soil micro-organisms
- Unexpected characteristics through genetic variability
- Mixed virus infections
- New pests and diseases

Socio-economic Issues and Ethical Consideration

One of the many questions being asked is how the GM technology is going to affect the incomes and livelihoods of the poor resource small farmers, who are the majority populations in the developing countries. The answer differs depending on the appropriateness, origin, ownership and control of the GM crops in question. Most of the GM technology is in the hands of the Transnational Companies (TNCs), who have developed most of the GM crops currently on the market, have put a stronger IPR protection to the technology and the crops. This raises fear of being controlled by 'foreign dominions'. The potential risks faced by the rural communities in developing countries are related to the:

- Monopoly control that the TNCs' developing country agents/subsidiaries/joint ventures exercise on the price of the GM seeds.
- Need to buy GM seeds for every new planting season to maintain high yield levels and fulfil farmers agreements with the seed-selling companies.
- Dependency on new generation of seeds or a reversion to old technology to address resistance that plant pests and diseases are likely to develop.
- Profit margins being squeezed between increasing seed prices and declining harvest selling prices.
- Possible loss of existing robust crop varieties and technologies, thereby reducing.
- The diversity, flexibility and resilience of farming systems and increasing vulnerability to events that could lead to famine.

These concerns are not unique to GM crops. The same situation was experienced when hybrid seed and elite cultivars were introduced some decades ago.

REVIEW QUESTIONS

1. What are the traditional and modern applications of biotechnology?
2. What are the various unit operations involved in downstream processing?
3. Justify the statement "Biotechnology is really an interdisciplinary subject".
4. What are the various steps outlined in integrated bioprocessing?
5. Distinguish between unit operation and unit process.
6. What is the role of biotechnology in society?
7. What is the role of biotechnology in environment cleanup?
8. Describe the applications of biotechnology in health and medicine.
9. Describe the applications of biotechnology in the field of agriculture.
10. What is a genetically modified organism? What impacts of GMOs on our environment?
11. Write a short note on Environmental safety issues.
12. Describe the Biosafety issues related with GMOs.

2

BIOMOLECULES

2.1 WHAT IS BIOMOLECULE?

Molecules containing covalently bonded carbon backbone are called organic compounds, they occur in an almost unlimited range. Mainly, biomolecules are organic compounds; we can therefore deduce that the bonding adaptability of carbon was a key factor in the selection of carbon compounds for the molecular machinery of cells during the origin and evolution of living organisms. Biomolecule includes large polymeric molecules such as carbohydrates, lipid, proteins, and nucleic acids. Most biomolecules can be regarded as derivatives of hydrocarbons and their functional groups determine its chemical property. In hydrocarbons, compounds with a covalently linked carbon backbone to which only hydrogen atoms are bonded. The backbones of hydrocarbons are very stable. Hydrogen atoms may be replaced by a variety of functional groups to yield different families of organic compounds. Typical families of organic compounds are the alcohols, which have one or more hydroxyl groups; amines, which have amino groups; aldehydes and ketones, which have carbonyl group; and carboxylic acids, which have carboxyl groups. Many biomolecules are polyfunctional, containing two or more different kinds of functional groups, each with its own chemical characteristics and reactions. Biomolecules almost never exist within cells as crystals rather; they are dissolved in the cytosol or associated with some other components of the cell. The important biomolecules (macromolecules) with their respective building blocks and major functions are given in Table 2.1.

Table 2.1 The major complex biomolecules of cells

S.No.	Biomolecule	Building block	Major functions
1.	Carbohydrates	Monosaccharides (Aldoses and Ketoses)	Storage form of energy to meet short term demands
2.	Protein	Amino acids	Fundamental basis of structure and function of cell
3.	Lipids	Fatty acids, Glycerol	Storage form of energy to meet long term demands, structural components of membrane
4.	DNA	Deoxyribonucleotides	Repository of hereditary information
5.	RNA	Ribonucleotides	Essentially required for protein biosynthesis

2.2 STRUCTURE AND FUNCTIONS OF BIOMOLECULES (MACROMOLECULES)

2.2.1 Carbohydrates

Carbohydrates or saccharides (Greek: *sakcharon*, sugar) are a widely diverse group of compounds that are ubiquitous in nature and very essential components of all living organisms and are, infect, the most abundant class of biomolecules. Literally, carbohydrate means "carbon hydrate" by their chemical composition, which is roughly, contain carbon, hydrogen and oxygen. The basic building blocks of carbohydrates are known as Monosaccharides. Typical carbohydrates are composed of strings or chains of monosaccharides *i.e.*, chains of individual sugars. A monosaccharide (mono = one, saccharide = sugar) is the smallest carbohydrate unit. The type of monosaccharides in the chain, length of the chain, and method of linking all determine the composition of the carbohydrate.

Classification of Carbohydrates

Carbohydrates are divided in to four major groups as follows:

I. Monosaccharides
II. Disaccharides
III. Oligosaccharides
IV. Polysaccharides

I. Monosaccharides

Monosaccharides are simple sugars with potentially free aldehyde or ketone groups. The simplest class of the carbohydrates are the monosaccharides, can be linked by glycosidic bonds to create larger units of disaccharides, oligosaccharides and polysaccharides. Monosaccharides are colourless, crystalline solids that are freely soluble in water, but insoluble in non polar solvents. Most of the monosaccharides have a sweet taste. The smallest Monosaccharides, those with three carbon atoms are trioses. Those with four, five, six, seven etc, C atoms are, respectively, tetroses, pentoses, hexoses, heptoses, etc. (as shown in Table 2.2). The general formula of monosaccharide is $C_n(H_2O)_n$.

Table 2.2

S.No.	Carbon atoms	Aldoses	Ketoses
1.	Trioses	Glyceraldehyde or Glycerose	Dihydroxy acetone
2.	Tetroses	Erythrose	Erythrulose
3.	Pentoses	Ribose, Xylose, Arabinose	Ribulose, Xylulose
4.	Hexoses	Galactose, Glucose, Mannose	Fructose
5.	Heptoses	Glucoheptose and Galactoheptose	Sedoheptulose

Trioses and Tetroses of Physiologic Importance

Phosphate of the two trioses, glyceraldehyde and Dihydroxy acetone, are formed from fructose-1, 6-diphosphate by glycolysis. Three phosphoglyceraldehyde and Erythrose-4-phosphate are formed by the hexose monophosphate shunt.

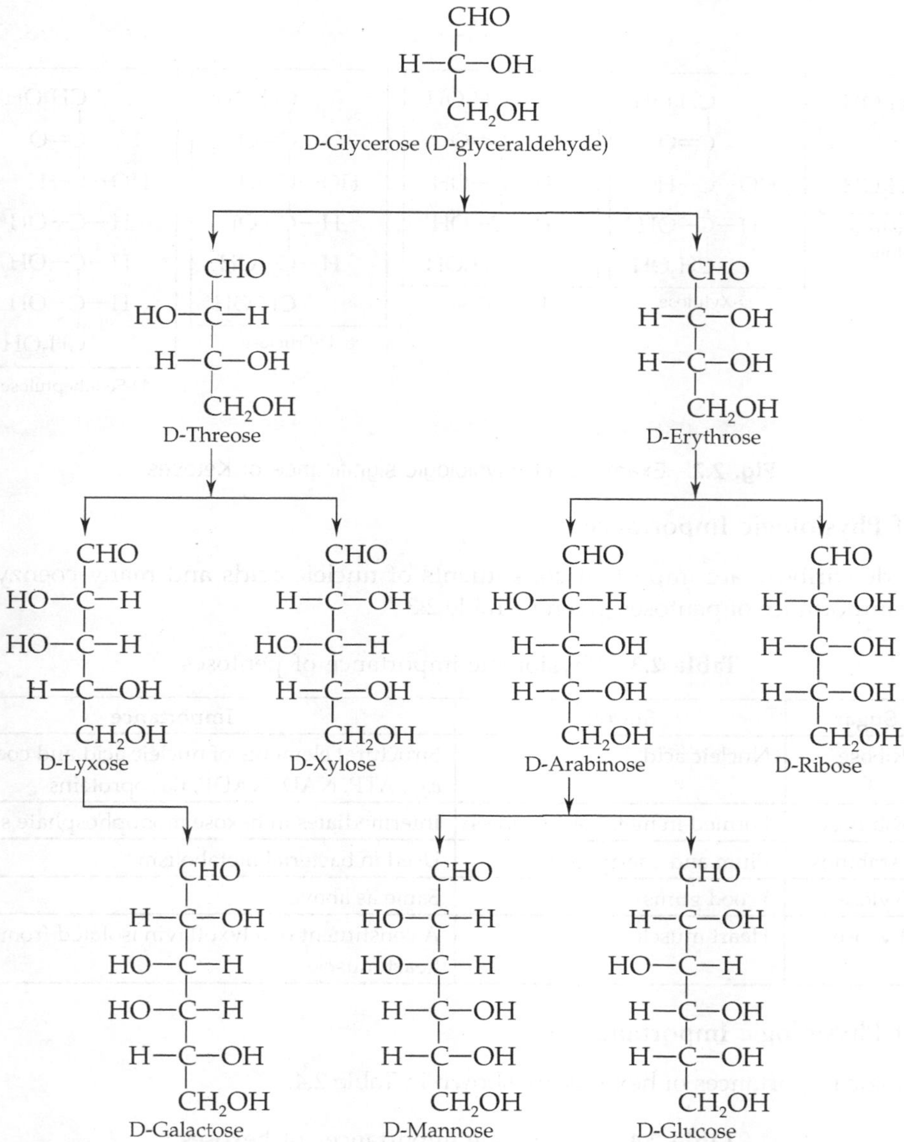

Fig. 2.1 Structural relations of the Aldoses, D-series of physiologic significance

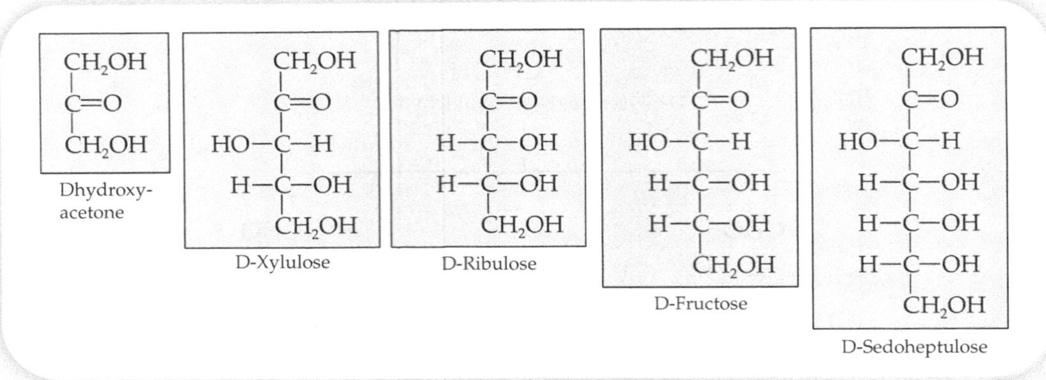

Fig. 2.2 Examples of physiologic significance of Ketoses

Pentoses of Physiologic Importance

Ribose and deoxiribose are important constituents of nucleic acids and many coenzymes. The physiologic importance of pentoses given in Table 2.3.

Table 2.3 Physiologic importance of pentoses

S.No.	Sugar	Source	Importance
1.	D-Ribose	Nucleic acid	Structural elements of nucleic acid and coenzymes, *e.g.*, ATP, NAD, NADP, flavoproteins
2.	D-Ribulose	Formed in metabolic process	Intermediates in hexose monophosphate shunt
3.	D-Arabinose	Plum and cherry gums	Used in bacterial metabolism
4.	D-Xylose	Wood gums	Same as above
5.	D-Lyxose	Heart muscle	A constituent of a lyxoflavin isolated from human heart muscle

Hexoses of Physiologic Importance

The physiologic importances of hexoses are shown in Table 2.4.

Table 2.4 Physiologic importances of hexoses

S.No.	Sugar	Source	Importance
1.	D-Glucose	Fruit juices, hydrolysis of starch, cane sugar, maltose and lactose	It is carried by the blood and used by the tissues
2.	D-Fructose	Fruit juice, honey, Hydrolysis of cane sugar and Inulin	It can be changed to glucose in the liver and intestine and thus used in the body
3.	D-Galactose	Hydrolysis of lactose	It can be changed to glucose in the liver and metabolized. Synthesized in the mammary gland to form lactose of milk. It is constituent of glycolipids and glycoproteins.
4.	D-mannose	Hydrolysis of plant mannosana and gums	It is the constituent of prosthetic polysaccharides of albumins, globulins, mucoproteins. It frequently occurs in glycoproteins.

Well Known Example of Hexose is Glucose

Glucose molecule occurs in two identical image (stereoisomer) forms, only one of which, the right-handed form (D-glucose). D-glucose form plays a major role in biology. D-glucose is frequently referred to as dextrose monohydrate, especially in the food industry. D-glucose is the only aldose that commonly occurs in nature as a monosaccharide.

Structure of D-Glucose

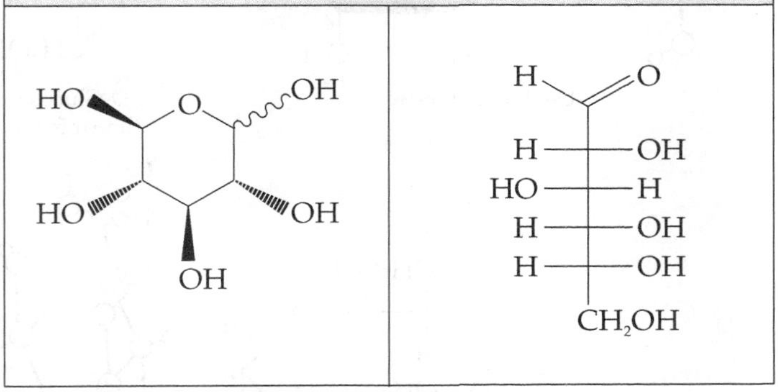

Fig. 2.3 Structure of Glucose

Isomerism

The presence of asymmetric carbon atoms allows the formation of isomers. The compounds which have the same structural formula but differ only in spatial configuration are called stereoisomers or geometric isomers. The α and β isomers of D-glucose in aqueous solution undergo optical rotation, reaching an equilibrium state in time. This is called mutarotation. It has been suggested that α and β isomers are not open chain structures, but are ring forms generated by reaction of the OH groups at C-5 and with CHO group at C-1. Haworth in 1929 proposed that ring form (hemiacetals) of sugars containing an asymmetric carbon atom can exist in α and β anomers. Haworth formulas indicate arrangement of carbon atoms in space in ring forms. Pyranose rings exhibit two configurations, the chair form and the boat form. A few important sugars are also shown in Haworth projection in Figs. 2.4 and 2.5.

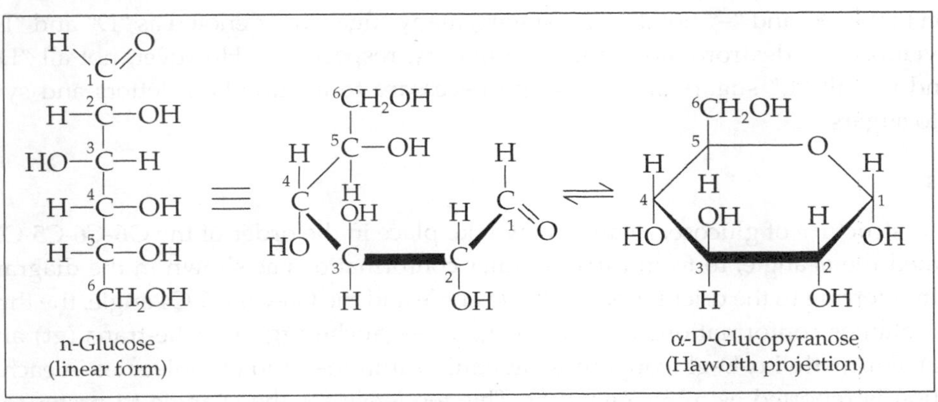

Fig. 2.4 The reaction of D-glucose to yield the cyclic hemiacetal α-D-glucopyranose

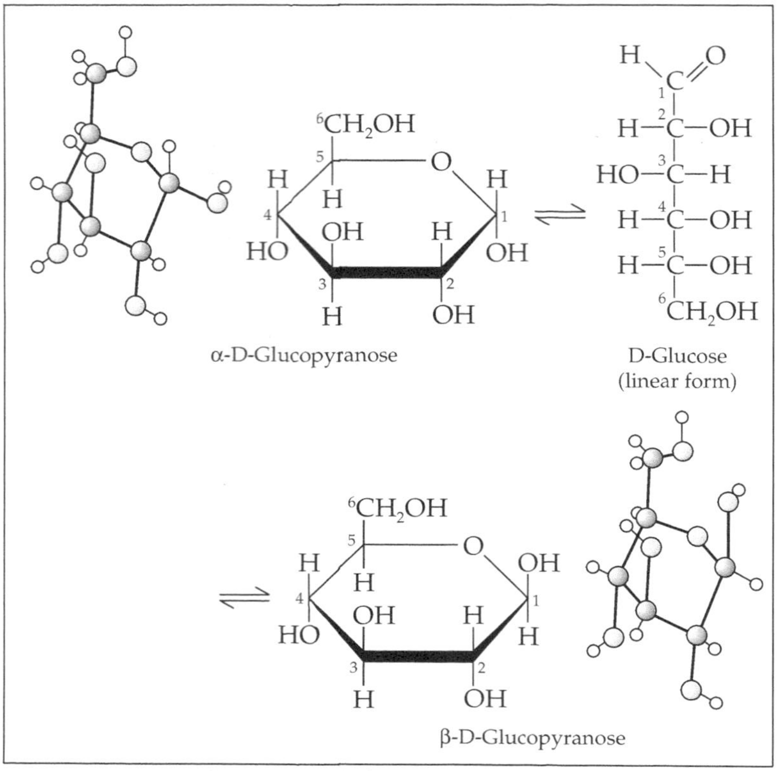

Fig. 2.5 Anomeric monosaccharides α-D-glucopyranose and β-D-glucopyranose

Optical Isomer

An important consideration in understanding sugars is that the sugars can bend (or rotate) a source of polarized light in one of two directions—when the light is bent in a clockwise (or right) direction (dextrorotatory) and in a counter clockwise (left) direction (levorotatory). Dextrorotatory substances are often denoted by a + sign, and levorotatory substances are often denoted by a – sign. The model of lactic acid (Fig. 2.6) is an example of how these two otherwise similar molecules exist. The situation is similar to two gloves – one for the right hand and the other for the left hand. In addition to the "+" and "–" notation for sugars, many sugars are denoted as "D" and "L", which are abbreviations for dextrorotatory and levorotatory, respectively. However, not all "D" sugars are "+" and not all "L" sugars are "–", so it's necessary to include both letters and symbols in referring to sugars.

Rotamers

In the cyclic structure of glucose, rotation may take place in the order of the O6-C6-C5-O5 torsion angle, termed the ω-angle, to form three rotamer conformations as shown in the diagram below (Fig. 2.7). In referring to the orientations of the ω-angle and the O6-C6-C5-C4 angle, the three stable staggered rotamer conformations are termed gauche-gauche (gg), gauche-trans (gt) and trans-gauche (tg). For methyl α-D-glucopyranose at equilibrium the ratio of molecules in each rotamer conformation is reported as 57:38:5 gg:gt:tg. This tendency for the ω-angle to prefer to adopt a gauche conformation is attributed to the gauche effect.

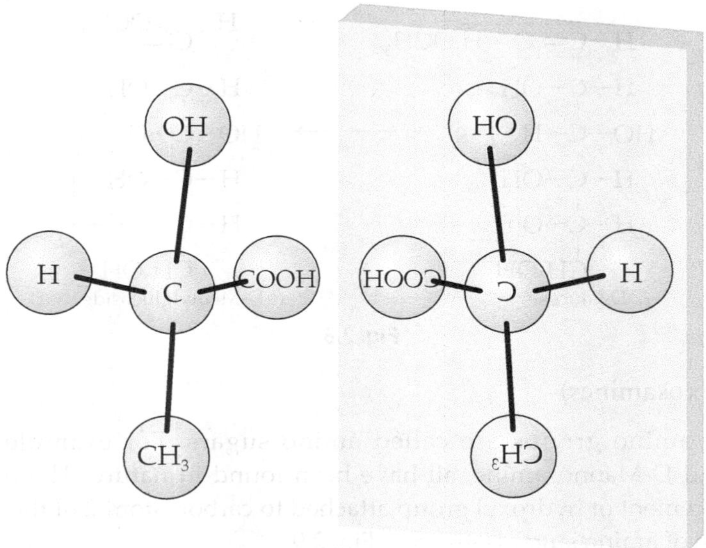

Fig. 2.6 Example of two isomers of lactic acid that exist as mirror image of each other

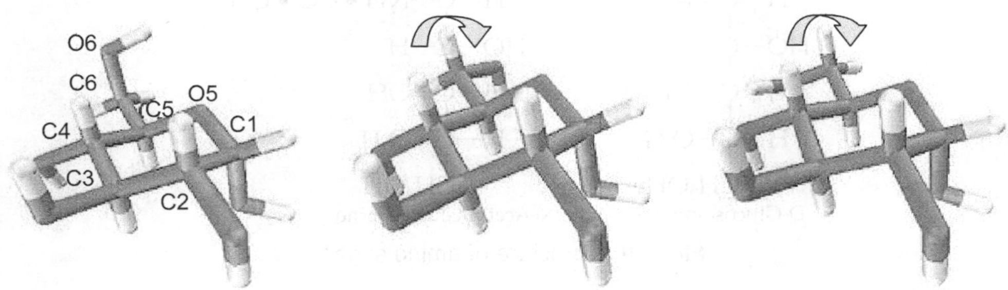

Fig. 2.7 Rotamer conformations of α-D-glucopyranose

Heptoses

Sedoheptulose is a ketoheptose found in plants. Its phosphate is important as an intermediate in the hexose monophosphate shunt and has been identified as a product of photosynthesis.

Glycosides

Glycosides are compounds formed by the condensation reaction between a sugar and the hydroxyl group of a second compound or aglycone which may or may not be an other sugar. If the carbohydrate portion is glucose, the resulting compound is glucoside: if galactose, a galactoside etc., all reducing sugars will condense with dry methyl alcohol under catalytic action of dry HCL to form glycosides. Thus, glucose forms α and β methyl glucosides. The aglycone may be –OH containing substances such as methyl alcohol, glycerol, a sterol, a phenol or another sugar.

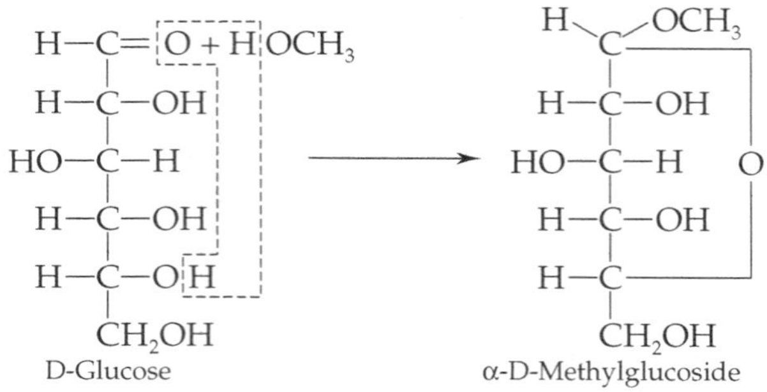

Fig. 2.8

Amino sugars (Hexosamines)

Sugars containing amino groups are called amino sugars. For example, D-glucosamine, D-galactosamine and D-Mannosamine, all have been found in nature. These amino sugars are formed by the replacement of hydroxyl group attached to carbon atom-2 of the sugar by an amino group. The structure of amino sugar is given in Fig. 2.9.

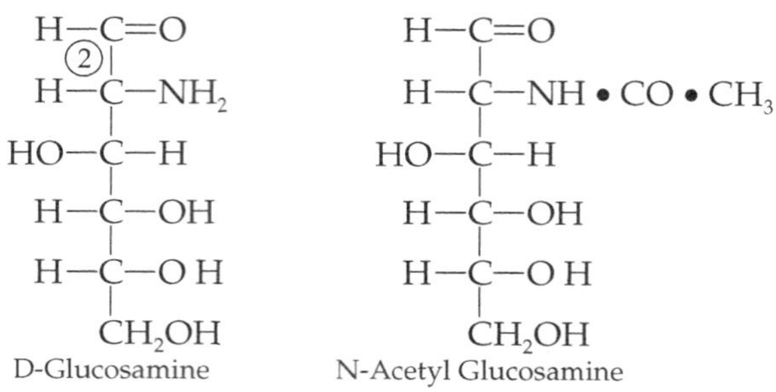

Fig. 2.9 Structure of amino sugar

II. Disaccharides

The disaccharides are composed of two monosaccharide units united by a glycosidic linkage.

- Sucrose – glucose + fructose
- Maltose – glucose + glucose
- Lactose – glucose + galactose

Sucrose

It is the most abundant disaccharide, occurs throughout the plant kingdom and is familiar to us as common table sugar. It is hydrolyzed to glucose and fructose by the enzyme invertase in the alimentary canal. It has no free aldehyde or keto group, because the linkage is between the aldehyde group of glucose and keto group of fructose. Since, it is a non reducing sugar it does not reduce Fehling's or Benedict's solutions.

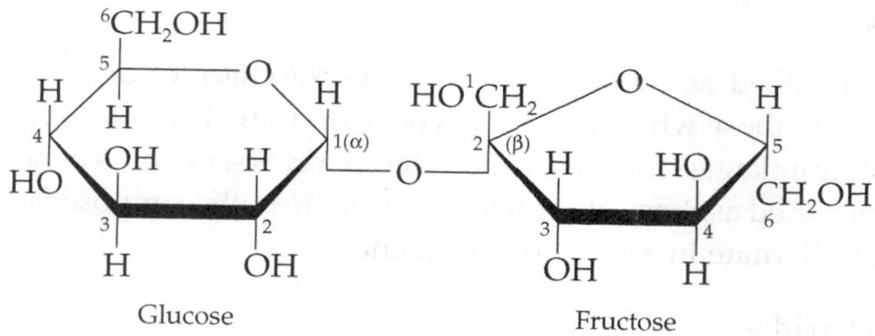

Sucrose

Fig. 2.10 Structure of Sucrose

Maltose

The sources of this are germinating cereal and malt and it is the intermediate product in the breakdown of starch by amylase in the alimentary canal. It is hydrolyzed to glucose by the enzyme maltase and the products are absorbed. It can reduce Fehling's or Benedict's solutions but cannot reduce Barfoed's solutions.

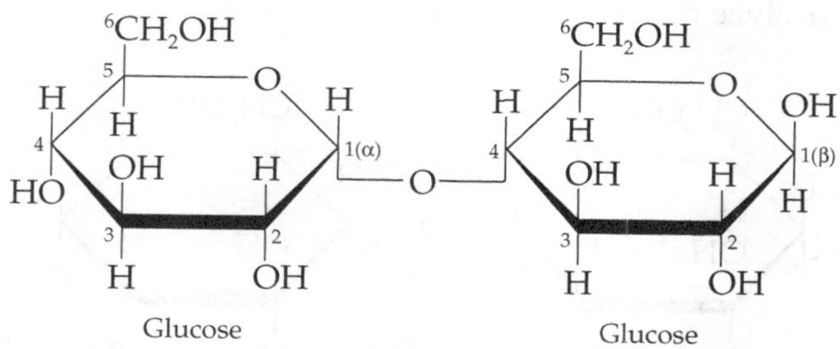

Maltose

Fig. 2.11 Structure of Maltose

Lactose

It is present in milk and formed in the lactating mammary gland. It may occur in urine during pregnancy. It is hydrolyzed to glucose and galactose by the enzyme lactase in the alimentary canal. It can reduce Fehling's or Benedict's solutions, but cannot reduce Barfoed's solutions.

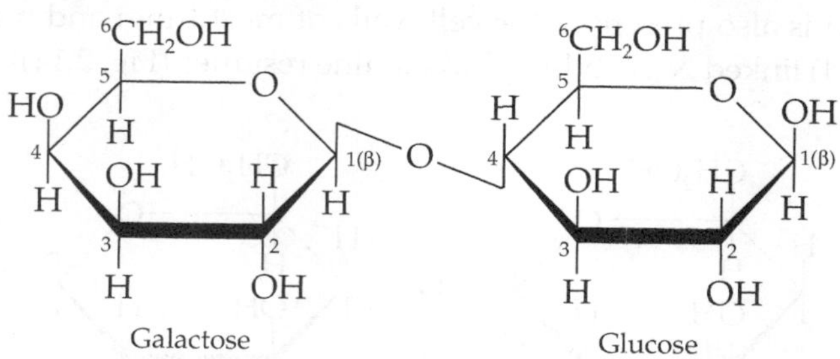

Lactose

Fig. 2.12 Structure of Lactose

III. Oligosaccharides

These carbohydrate yield 2–10 monosaccharide units on hydrolysis, for example Maltotriose, which contain three α(1→4) linked glucose residues. Some oligosaccharides are also known as dextrins that contain α(1→6) branches. These oligosaccharides are hydrolyzed to their component monosaccharides by specific enzyme contained in the brush border membranes of the intestinal mucosa.

IV. Polysaccharides

Polysaccharides are classified as two classes: homopolysaccharides and heteropolysaccharides. Homopolysaccharides are those which have one type of or more than one type of monosaccharides unit linked by glucosidic linkages. Although, monosaccharides sequences of heteropolysaccharides can, in principle, be as varied as those of proteins, they are usually composed of only a few types of monosaccharides that alternate in a repetitive sequence.

Structural Polysaccharides

Cellulose

Cellulose is the primary structural component of plants. It is found primarily in the cells walls and is a primary component of the fiber component of animal feeds. The structure of cellulose is similar to that of amylose in starch – that is, linear chains of glucose, except that the glucose molecules are joined by β-1,4-glucosidic linkages. This linkage is the reason that humans (and preruminant calves) cannot digest cellulose. Cattle digest cellulose, because rumen bacterias contain the enzyme cellulase, which can hydrolyze this linkage.

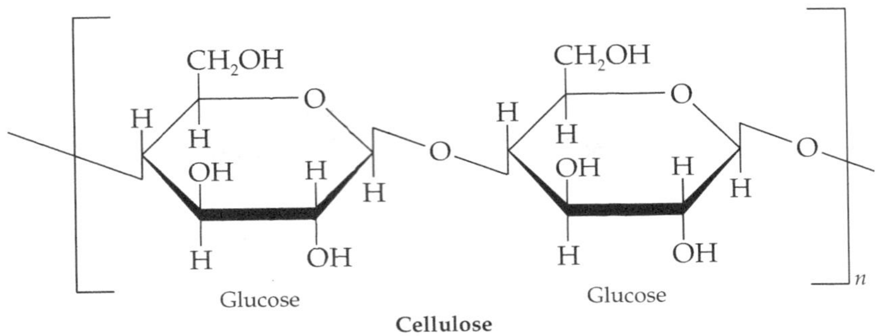

Fig. 2.13 Structure of Cellulose

Chitin

Chitin is a principle structural component of the exoskeletons of invertebrates such as crustaceans, insects and spiders and is also present in the cell walls of most fungi and many algae. Chitin is a homopolymer of β(1→4) linked N-acetyl-D-glucosamine residues (Fig. 2.14).

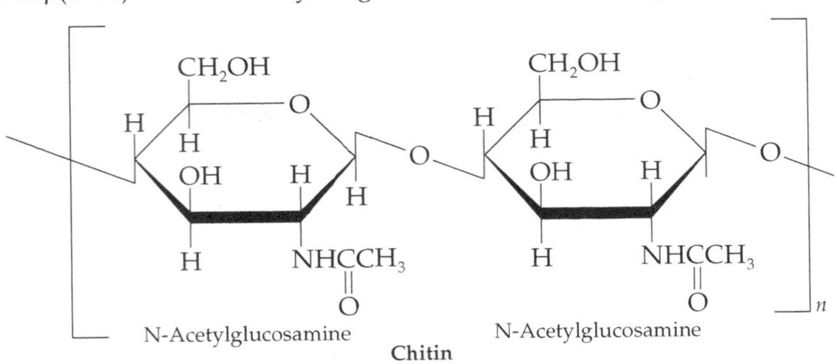

Fig. 2.14 Structure of Chitin

Storage Polysaccharides

Starch is a mixture of glucans that plant synthesized as their principal food reserve. It is deposited in the cytoplasm of plant cells as insoluble granules composed of α-Amylose and amylopectin. α-Amylose is a linear polymer of several thousand glucose residues linked by α(1→4) bonds (Fig. 2.15 (*a*)). Amylopectin (Fig. 2.15 (*b*)) consists mainly of α(1→4) linked glucose residues, but is a branched molecule with α(1→6) branch points every 24 to 30 glucose residues on average. Amylopectin molecules contain up to 10^6 glucose residues, which make them among the largest molecule occurring in nature.

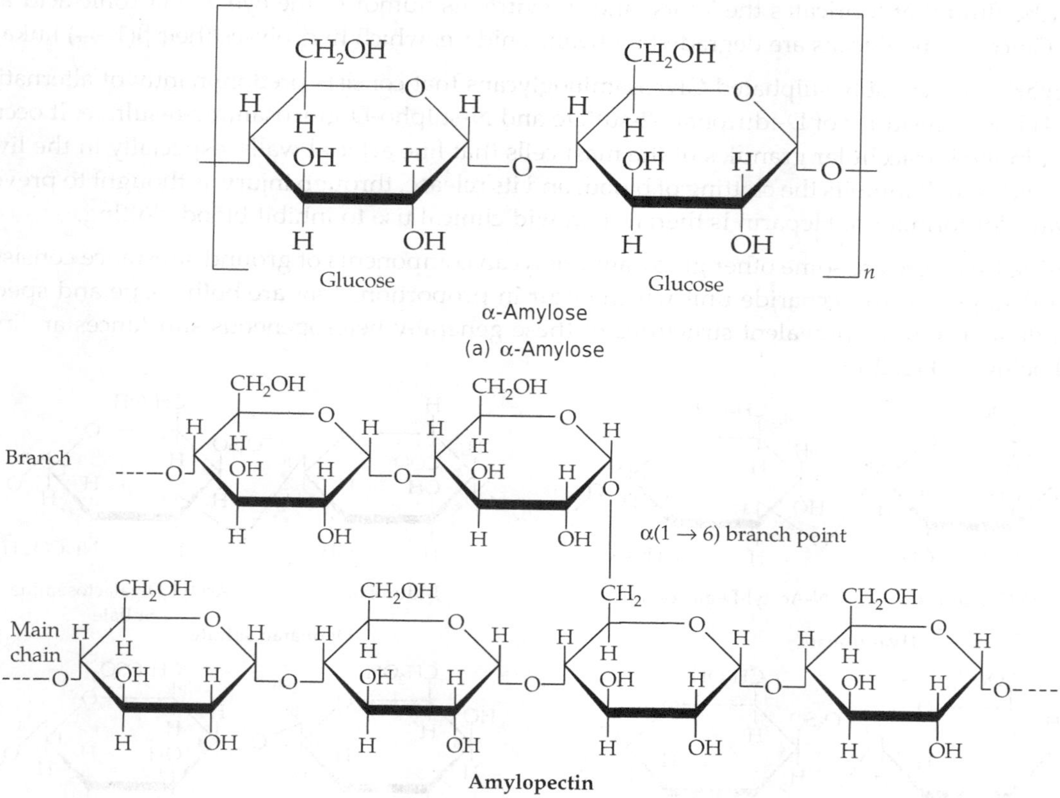

(b) Amylopectin

Fig. 2.15 Structure of (a) α-amylose and (b) amylopectin

Glycogen

Glycogen is a three dimensional strings of glucose that is produced by the liver and stored in liver and muscles. Glycogen serves as a glucose reserve for animals. It has a branched structure with straight chain units of 12–18, α-D-glucopyranose (in a α(1→4) glucosidic linkage) with branching by means of a α(1→6) glucosidic branch. In the cell, glycogen is degraded for metabolic use by glycogen phosphorylase, which phosphorolytically cleaves glycogen's α(1→4) bonds sequentially inward from its non reducing end to yield glucose-1-phosphate. Glycogen is a very highly branched structure, which has many non reducing ends, permits the rapid mobilization of glucose in times of metabolic need.

Inulin

It is found in tubers and roots of dahlias and dandelions. It is hydrolyzed to fructose hence, it is fructosan. It is easily soluble in warm water.

Glycosaminoglycans

The extra cellular spaces, particularly those of connective tissues such as cartilage, tendon, skin and blood vessel walls consists of collagen and elastin fibers embedded in a gel like matrix known as ground substance. Ground substance is composed largely of Glycosaminoglycans (mucopolysaccharides), unbranched polysaccharides of alternating uronic acid and Hexosamine residues.

Hyaluronic acid is an important Glycosaminoglycans component of ground substance, synovial fluid (the fluid that lubricates the joints) and the vitreous humor of the eye. Hyaluronic acid and other Glycosaminoglycans are degraded by hyaluronidase, which hydrolyses their β(1→4) linkage.

Heparin is a variably sulphated Glycosaminoglycans that consists predominantly of alternating α(1→4) linked residues of D-iduronate-2-sulfate and N-sulpho-D-glucosamine-6-sulfate. It occurs almost in the intracellular granules of the mast cells that line arterial walls, especially in the liver, lungs and skin. It inhibits the clotting of blood, and its release, through injury, is thought to prevent runway clot formation. Heparin is therefore in wid clinical use to inhibit blood clotting.

Beside this there are some other glycosaminoglycan components of ground substance consist of 50–1000 sulphated disaccharide unit which occur in proportions that are both tissue and species dependent. The most prevalent structures of these generally heterogeneous substances are indicated below in Fig. 2.16:

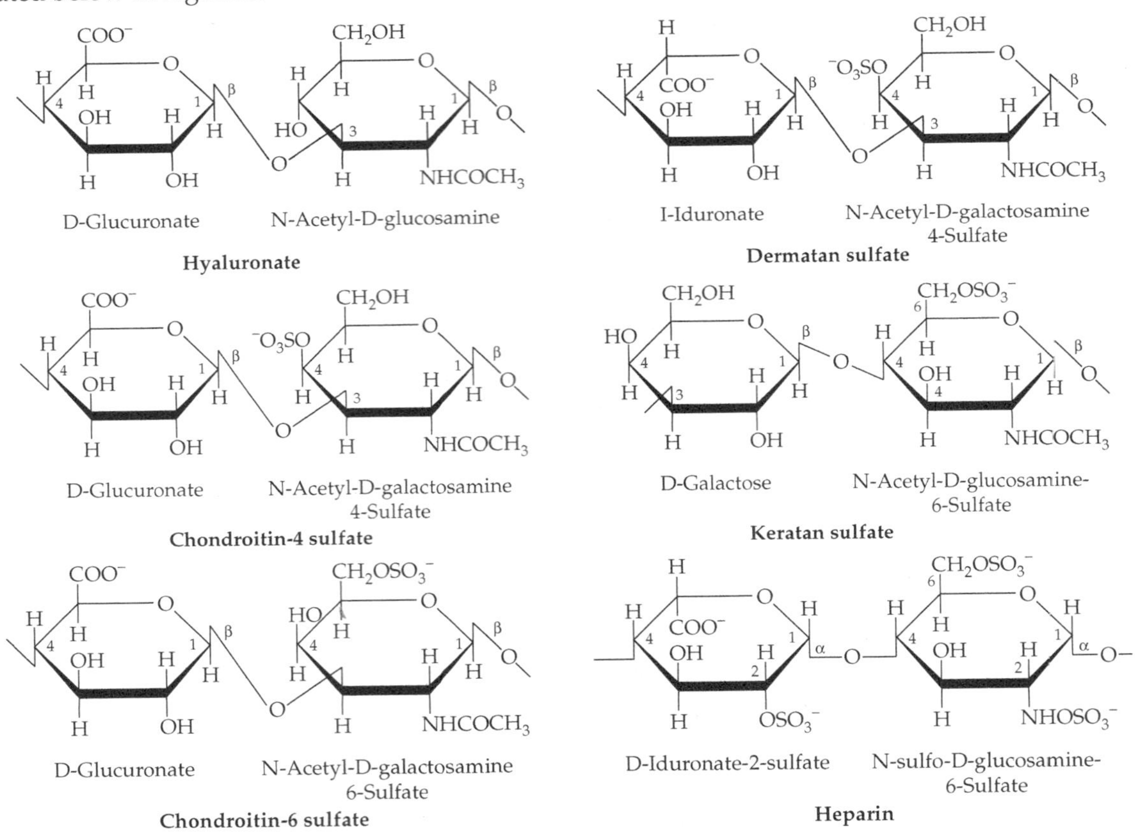

Fig. 2.16 The disaccharides repeating units of the common Glycosaminoglycans

Peptidoglycans (Bacterial cell wall)

The stiff constituent of bacterial cell walls is a heteropolymer of alternating β(1→4) linked N-acetylglucosamine and N-acetylmuramic acid residues. The linear polymers lie side by side in the cell wall, cross linked by short peptide. The enzyme lysozyme kills bacteria by hydrolysing the β(1→4) glycosidic bond between N-acetylglucosamine and N-acetylmuramic acid.

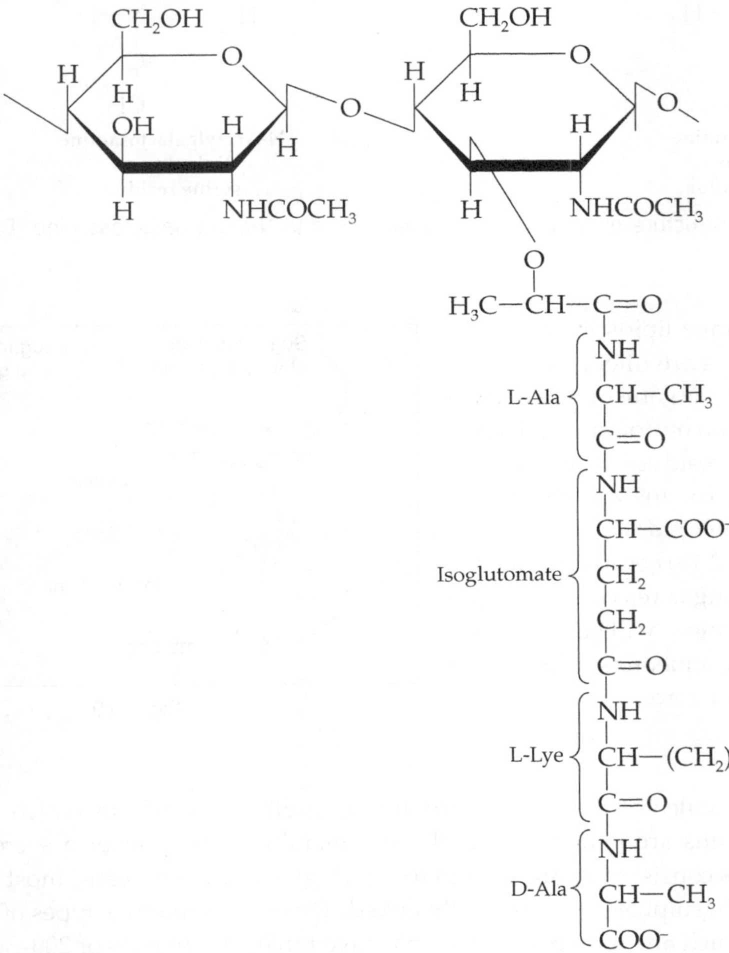

Fig. 2.17 The repeating unit of peptidolycan is an NAG-NAM

Specific type of Carbohydrate units positioned on the extra cellular surface of plasma membrane

Glycoprotein

Glycoproteins have one or several oligosaccharides of varying complexity joined covalently to a membrane protein. They are found on the outer face of the plasma membrane, extracellular matrix, and in the blood. In membrane glycoproteins, sugars are attached either to the amide nitrogen atom in the side chain of asparagines (termed N-linkage) or to the oxygen atom in the side chain of serine or threonine (termed an O-linkage). The sugar directly attached to one of these side chains is usually N-acetylglucosamine and N-acetylgalactosamine (Fig. 2.18 A and B).

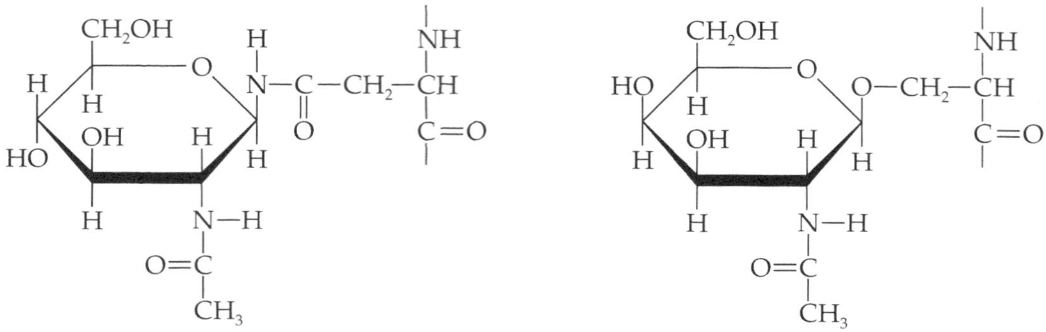

A N-Acetylgulcosamine linked to an asparagine residue

B N-Acetylgalactosamine linked to an serine residue

Fig. 2.18 Structure of N-acetylglucosamine (A) and N-acetylgalactosamine (B)

Glycolipids

Glycolipids are membrane lipids in which the hydrophilic head groups are oligosaccharides. The glycolipids of higher organisms are derivatives are sphingosine with one or more attached sugar residues. Sugar residues of membrane glycolipids and glycoproteins are always located on the extracellular side of mammalian plasma membrane (Fig. 2.19) because sugars are very hydrophilic, the sugar residues of a glycolipid are of a glycoprotein will strongly prefer to be located near the aqueous surface rather than in the hydrocarbon core.

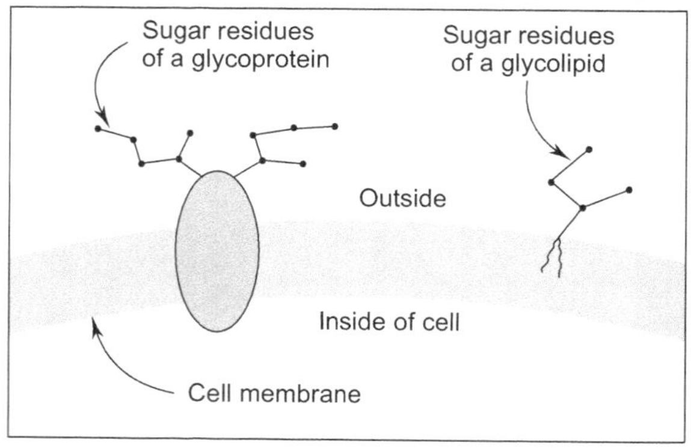

Fig. 2.19

Proteoglycans

Proteoglycans are molecules of the cell surface are extracellular matrix in which one or more glycosaminoglycan chains are joined covalently to a membrane protein or a secreted protein. Proteoglycans sub units consist of a core protein to which glycosaminoglycans, most often keratin sulphate and chondroitin sulphate, are covalently linked. There are numerous types of core protein, those from cartilage, which are known as aggrecans, have molecular masses of 200–300 kD, which makes them among the largest polypeptides chains synthesized by any cells.

Functions of Carbohydrates

Carbohydrates participate in a wide range of functions:

1. Carbohydrates are precursors for many organic compounds (amino acids, fat).
2. They are structural components of many organisms. These include the fiber (cellulose) of plants, exoskeleton of some insects and the cell wall of microorganisms.
3. They are the most abundant dietary source of energy for all organisms. Carbohydrates Supplies energy to body, brain and the nervous system.

4. Carbohydrates (as glycoprotein and glycolipids) participate in the structure of cell membrane and cellular functions such as cell growth, adhesion and fertilization.
5. Carbohydrates are utilized as raw materials for several industries for *e.g.*, Paper, plastic, textile, alcohol etc.
6. Carbohydrates also serve as the storage form of energy (glycogen) to meet the immediate energy demands of the body.

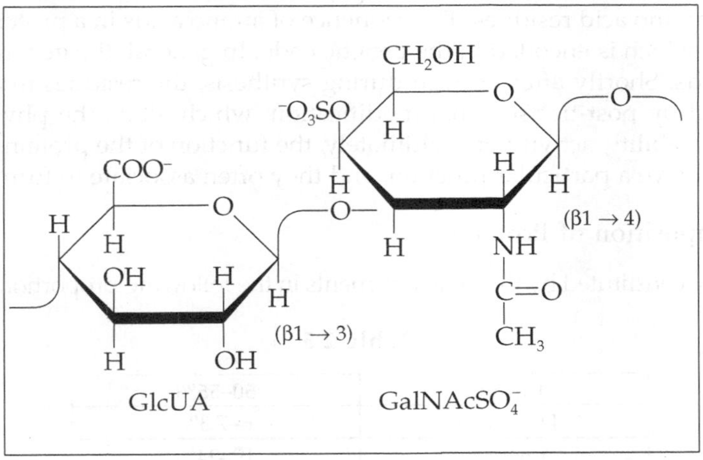

(a)

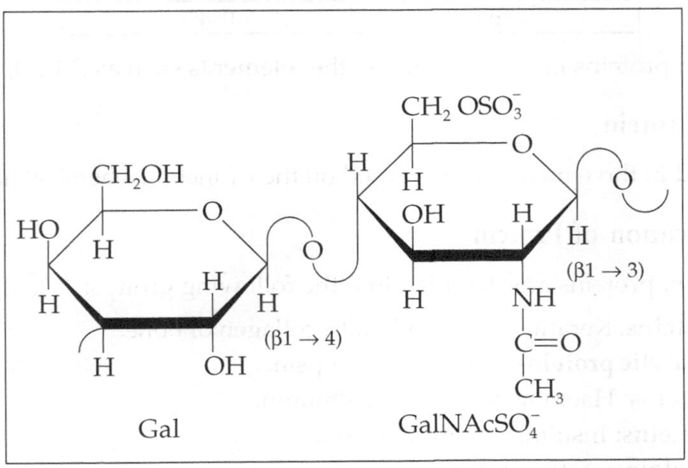

(b)

Fig. 2.20 Structure of Chondroitin sulphate (*a*) and Keratin sulphate (*b*)

Side Effects of Carbohydrates

Side effects of Carbohydrates Excessive carbohydrates can cause an increase in the total caloric intake, causing obesity and pancreatic cancer in women. Deficient carbohydrates can cause a lack of calories (malnutrition), or excessive intake of fats to make up the calories. Low-fat, high-carbohydrate diets contribute to hyperinsulinemia and hypertriglyceridemia. The diseases diabetes mellitus, galactosemia, glycogen storage diseases and milk intolerance are directly associated with carbohydrates.

2.2.2 Proteins

The term protein is derived from a *Greek word* **'proteios'**, meaning 'holding the first place.' Proteins are an important class of biological macromolecules present in all biological organisms, made up primarily from the elements carbon, hydrogen, nitrogen, oxygen, and sulphur. All proteins are polymers of amino acids, arranged in a linear chain and folded into a globular form. The amino acids in a polymer are joined together by the peptide bonds between the carboxyl and amino groups of adjacent amino acid residues. The sequence of amino acids in a protein is defined by the sequence of a gene, which is encoded in the genetic code. In general, the genetic code specifies 20 standard amino acids. Shortly after or even during synthesis, the residues in a protein are often chemically modified by post-translational modification, which alters the physical and chemical properties, folding, stability, activity, and ultimately, the function of the proteins. Proteins can also work together to achieve a particular function, and they often associate to form stable complexes.

Fundamental Composition of Proteins

Proteins are primarily constituted by five major elements in the following proportion shown in Table 2.5:

Table 2.5

Carbon	50–55%
Hydrogen	6–7.3%
Oxygen	19–24%
Nitrogen	13–19%
Sulpher	0–4%

Besides the above, proteins may also contain other elements such as P, Fe, I, Cu, Mn, Zn, etc.

Classifications of Protein

Proteins are classified in three major groups based on their function, chemical nature and solubility.

Functional Classification of Protein

Based on the function, proteins are classified into the following groups:

- **Structural proteins:** Keratin of hair and nails, collagen of bone.
- **Enzyme or catalytic proteins:** Hexokinase, pepsin.
- **Transport proteins:** Haemoglobin, serum albumin.
- **Hormonal proteins:** Insulin, growth hormone.
- **Contractile proteins:** Actin, myosin.
- **Storage proteins:** Ovalbumin, glutalin.
- **Genetic proteins:** Nucleoprotein.
- **Defence proteins:** Snake venoms, immunoglobulins.
- **Receptor proteins:** for hormones, viruses.

Proteins are classified on to basis of their chemical nature and solubility

This is more comprehensive and popular classification of proteins. It is based on the amino acid composition, structure, shape and solubility. Proteins are broadly classified into three groups as shown in Tables 2.6, 2.7 and 2.8.

Simple proteins: They are composed only amino acid residues.

Conjugated proteins: Besides the amino acids, these proteins contain a non-protein moiety known as prosthetic group or conjugated group.

Derived proteins: These are the denatured or degraded products of simple and conjugated proteins.

Table 2.6 Simple proteins

Class of proteins	Characteristics	Examples
(*i*) Albumins	Soluble in water, coagulable by heat and precipitated at high salt concentrations.	Serum albumin, egg albumin, lactalbumin (Milk), leucosin (wheat),, legumelin (soyabeans).
(*ii*) Globulins	Insoluble in water, soluble in dilute salt solutions and precipitated by half saturated salt solutions.	Serum globulin, vitellin (egg yolk), tuberin (potato), myosinogen (muscle), legumin (peas).
(*iii*) Glutelins	Insoluble in water but soluble in dilute acids and alkalis. Mostly found in plants.	Glutenin (wheat), oryzenin (rice).
(*iv*) Prolamines	Insoluble in water and absolute alcohol but soluble in 70 to 80 per cent alcohol.	Gliadin (wheat), zein (maize).
(*v*) Protamines	Basic proteins of low molecular weight. Soluble in water, dilute acids and alkalis. Not coagulable by heat.	Salmine (salmon sperm).
(*vi*) Histones	Soluble in water and insoluble in very dilute ammonium hydroxide.	Globin of hemoglobin and thymus his-tones.
(*vii*) Scleroproteins	Insoluble in water, dilute acids and alkalis.	Keratin (hair, horn, nail, hoof and feathers), collagen (bone, skin), elastin (ligament).

Table 2.7 Conjugated proteins

Class of proteins	Characteristics	Examples
(*i*) Nucleoproteins	Composed of simple basic proteins (protamines or histones) with nucleic acids, found in nuclei. Soluble in water.	Nucleoprotamines and nucleohistones.
(*ii*) Lipoproteins	Combination of proteins with lipids, such as fatty acids, cholesterol and phospholipids etc.	Lipoproteins of egg-yolk, milk and cell membrances, lipoproteins of blood.
(*iii*) Glycoproteins	Combination of proteins with car-bohydrate (Mucopolysaccharides).	Mucin (saliva), ovomucoid (egg white), osseomucoid (bone), tendomucoid (tendon).
(*iv*) Phosphoproteins	Contain phosphorus radical as a prosthetic group	Caseinogen (milk), ovovitellin (egg-yolk).
(*v*) Metalloproteins	Contain metal ions as their prosthetic groups. The metal ions generally are Fe, Co, Mg, Mn, Zn, Cu etc.	Siderophilin (Fe), ceruloplasmin (Cu).
(*vi*) Chromoproteins	Contain porphyrin (with a metal ion) as their prosthetic groups.	Haemoglobin, myoglobin, catalase, peroxidase, cytochromes.
(*vii*) Flavoproteins	Contain riboflavin as their prosthetic groups.	Flavoproteins of liver and kidney.

Table 2.8 Derived proteins

Class of proteins	Characteristics	Examples
A. *Primary derivatives :*		
(*i*) Proteans	Derived in the early stage of protein hydrolysis by dilute acids, enzymes or alkalis.	Fibrin from fibrinogen.
(*ii*) Metaproteins	Derived in the later stage of protein hydrolysis by slightly stronger acids and alkalis.	Acid and alkali metaproteins.
(*iii*) Coagulated proteins	They are denatured proteins formed by the action of heat, X-rays, ultraviolet rays etc.	Cooked proteins, coagulated albumins.
B. *Secondary derivatives :* (Due to cleavage of peptide bonds)		
(*i*) Proteoses	Formed by the action of pepsin or trypsin. Precipitated by saturated solution of ammonium sulphate, incoagulable by heat.	Albumose from albumin, globulose from globulin.
(*ii*) Peptones	Further stage of cleavage than the proteoses. Soluble in water, incoagulable by heat and not precipitated by saturated ammonium sulphate solutions.	
(*iii*) Peptides	Compounds containing two or more amino acids. They may be di-, tri-, and polypeptides.	Glycyl-alanine, leucyl-glutamic acid.

Building Blocks of Protein

Proteins are the polypeptide chains of amino acid. Amino acids are the building block of protein. As many as 300 amino acids occur in nature: of these, only 20 known amino acids are repeatedly found in the structure of proteins.

Amino Acids

Amino acids are molecules containing an amino group ($-NH_2$) and carboxylic acid group ($-COOH$) and a side chain that varies between different amino acids. These molecules contain the key elements of Carbon, Hydrogen, Oxygen, and Nitrogen. These molecules are particularly important in biochemistry, where this term refers to alpha-amino acids with the general formula $H_2NCHRCOOH$, where R is an organic substituent. In an alpha amino acid, the amino and carboxylate groups are attached to the same carbon atom, which is called the α-carbon. The various alpha amino acids differ in which side chain (R-group) is attached to their alpha carbon. These side chains can vary in size from just a hydrogen atom in glycine, to a methyl group in alanine, through to a large heterocyclic group in tryptophan.

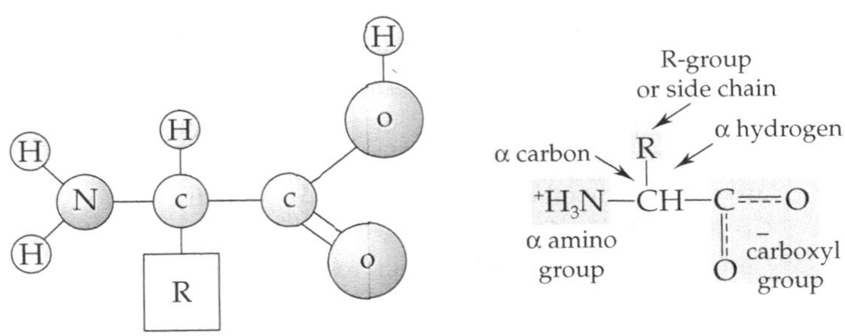

Fig. 2.21 The Structure of a typical amino acid showing α-carbon and R-groups

Classification of Standard Amino Acids

There are different ways of classifying the amino acids based on the structure and chemical nature, nutritional requirement and metabolic fate.

(*a*) **Classification of amino acids based on the structure :** A comprehensive classification of amino acids is based on the structure. Each amino acid is assigned a 3 letter or 1 letter symbol. These symbols are commonly used to represent the amino acids. The 20 amino acids found in proteins are divided in to 7 distinct groups as shown in Table 2.9.

(*b*) **Classification of amino acids based on the polarity :** Amino acids are classified in to 3 groups based on their polarity, as shown in Table 2.10.

(*i*) Non polar amino acids

(*ii*) Polar amino acids with uncharged polar side chain

(*iii*) Amino acids with charged polar side chain

(*c*) **Classification of amino acids based on the nutrition :** Based on the nutritional requirement, amino acids are grouped in to 2 classes shown in Table 2.11.

Essential amino acid : The amino acids which cannot be synthesized by the body and, therefore need to be supplied through the diet are called essential amino acids. They are required for proper growth and maintenance of the individual.

Non essential amino acid : The body can synthesize about ten amino acids to meet the biological needs; hence they need not be consumed in the diet.

(*d*) **Classification of amino acids based on their metabolic fate :** The carbon skeleton of amino acids can serve as a precursor for the synthesis of glucose or fat or both. From metabolic view point, amino acids are divided into three groups:

1. *Glycogenic amino acids*: These amino acids can serve as a precursor for the formation of glucose or glycogen. *e.g.*, alanine, aspartate, glycine, methionine etc.
2. *Ketogenic amino acids*: Fat can be synthesized from these amino acids. Two amino acids leucine and lysine are exclusively ketogenic.
3. *Glycogenic and ketogenic amino acids*: The four amino acids isoleucine, phenylalanine, tryptophan, tyrosine are precursors for synthesis of glucose as well as fat.

Non Standard Amino Acids

Moreover, the 20 standard amino acids (described above) present in the protein structure, there are several other amino acids which are biologically important. These include the amino acids derivatives found in proteins, non proteins amino acids performing specialized functions and the D-amino acids Fig. 2.22 (*a*) and (*b*).

(*a*) **Amino acids derivatives in proteins :** The 20 standard amino acids can be incorporated in to proteins due to the presence of universal genetic code. Some of these amino acids undergo specific modification after the protein synthesis occurs. These derivatives of amino acids are very important of protein structure and functions. Some of examples are given here:

(*i*) *Collagen*: The most abundant protein in mammals contains 4-hydroxyproline and 5-hydroxylysine

(*ii*) *Histones*: The proteins found in association with DNA, contain many methylated, phosporylated or acetylated amino acids.

(*iii*) γ-carboxyglutamic acid is found in certain plasma protein involved in blood clotting.

(*iv*) Cystine is formed by the combination of two cysteines. Cystine is also considered as derived amino acids.

(*b*) **Non protein amino acids :** These amino acids, although never found in proteins, perform several biologically important functions. They may be either α or non α amino acids. List of these amino acids along with their functions given in Table 2.12.

(*c*) **D-amino acids :** D-amino acids are found in the antibiotic (actinomycin-D, Valinomycin, gramicidin-S). D-glutamic acid and D-alanine are present in bacterial cell wall.

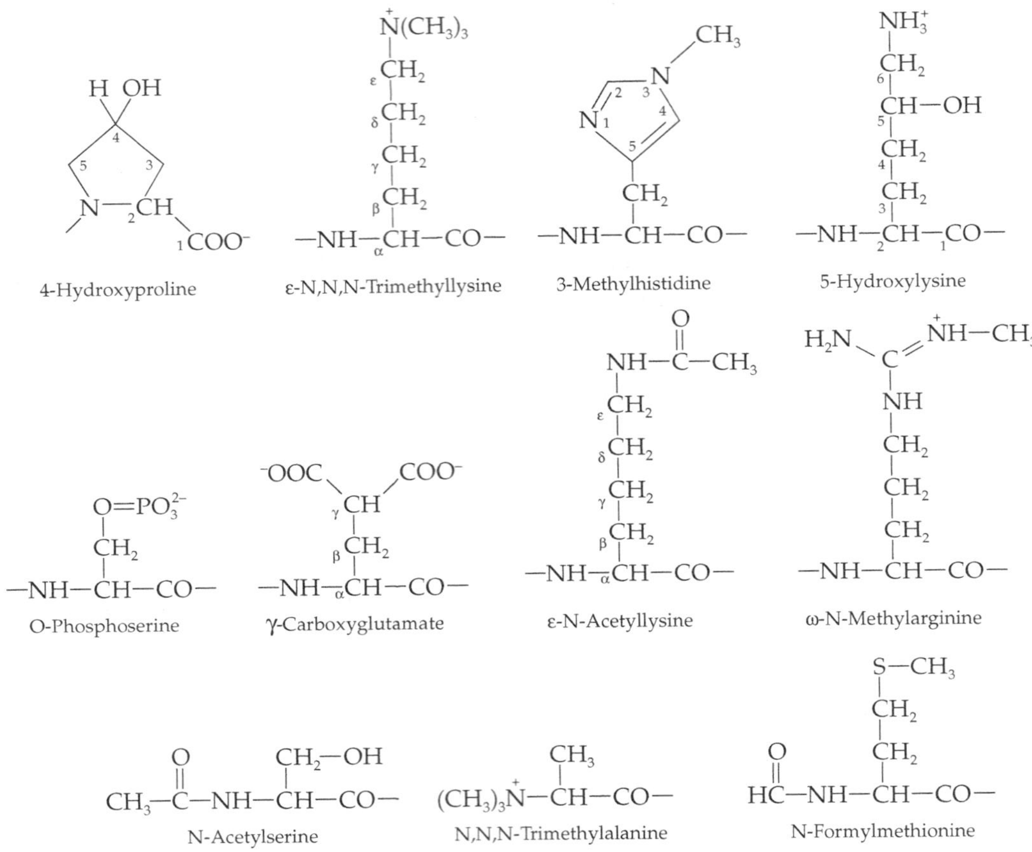

Fig. 2.22 (*a*) Some uncommon amino acid residues that are components of certain proteins. All of these residues are modified from one of the 20 standard amino acids after polypeptide chain biosynthesis

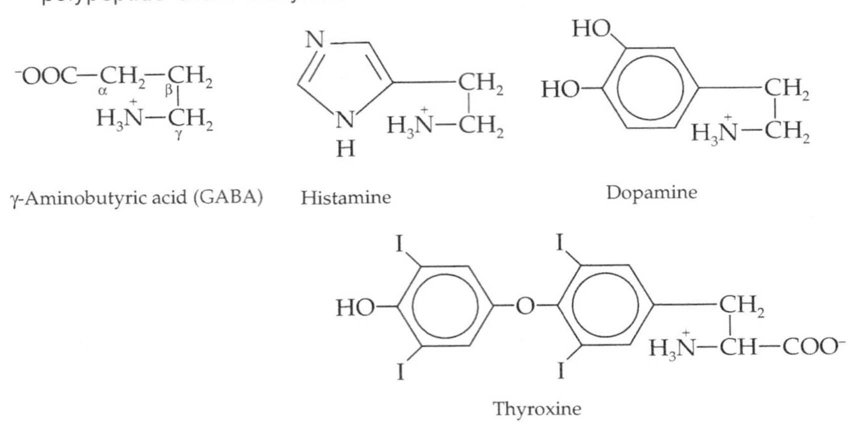

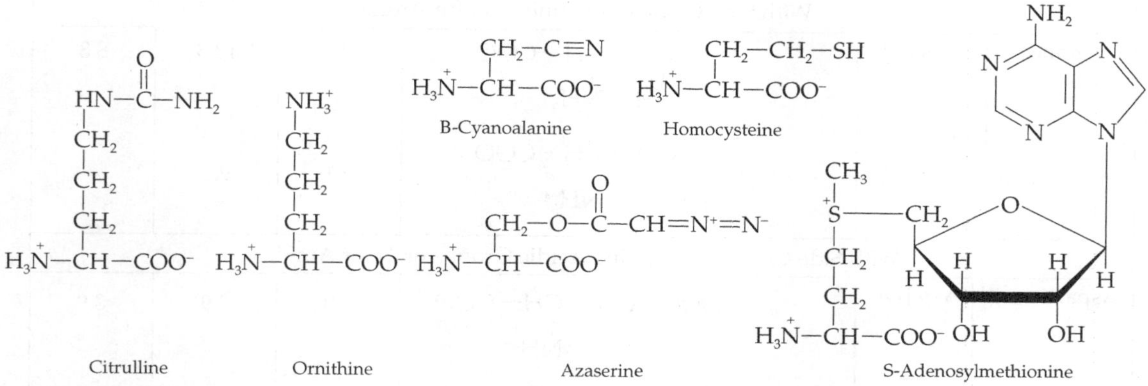

Fig. 2.22 (*b*) Some biologically produced derivatives of standard amino acids and amino acids that are not components of proteins

Table 2.9 Amino acids classification based on the structure

Name	Symbol	Structural Formula	pK_1	pK_2	pK_3
With Aliphatic Side Chains					
Glycine	Gly [G]	$H-CH(NH_3^+)-COO^-$	α-COOH 2.4	α-HN_3^+ 9.8	R Group
Alanine	Ala [A]	$CH_3-CH(NH_3^+)-COO^-$	2.4	9.9	
Valine	Val [V]	$(H_3C)_2CH-CH(NH_3^+)-COO^-$	2.2	9.7	
Leucine	Leu[L]	$(H_3C)_2CH-CH_2-CH(NH_3^+)-COO^-$	2.3	9.7	
Isoleucine	Ile[I]	$CH_3-CH_2-CH(CH_3)-CH(NH_3^+)-COO^-$	2.3	9.8	
With Side Chains Containing Hydroxylic (OH) Groups					
Serine	Ser [S]	$CH_2(OH)-CH(NH_3^+)-COO^-$	2.2	9.2	about 13
Threonine	Thr [T]				
Tyrosine	Tyt [Y]	$CH_3-CH(OH)-CH(NH_3^+)-COO^-$ See below	2.1	9.3	about 13

With Side Chains Containing Sulfur Atoms					
Cysteine	Cys [C]	$CH_2(SH)-CH(NH_3^+)-COO^-$	1.9	10.8	8.3
Methionine	Met [M]	$CH_2(S-CH_3)-CH_2-CH(NH_3^+)-COO^-$	2.1	9.3	
With Side Chains Containing Acidic Groups or Their Amides					
Aspartic acid	Asp [D]	$^-OOC-CH_2-CH(NH_3^+)-COO^-$	2.0	9.9	3.9
Asparagine	Asn [N]	$H_2N-C(=O)-CH_2-CH(NH_3^+)-COO^-$	2.1	8.8	
Glutamic acid	Glu [E]	$^-OOC-CH_2-CH_2-CH(NH_3^+)-COO^-$	2.1	9.5	4.1
Glutamine	Gln [Q]	$H_2N-C(=O)-CH_2-CH_2-CH(NH_3^+)-COO^-$	2.2	9.1	

Table 2.10

With Side Chains Containing Basic Groups					
			α-COOH	α-NH_3^+	R Group
Arginine	Arg [R]	$H-N(C(=NH_2^+)NH_2)-CH_2-CH_2-CH_2-CH(NH_3^+)-COO^-$	1.8	9.0	12.5
Lysine	Lys [K]	$CH_2(NH_3^+)-CH_2-CH_2-CH_2-CH(NH_3^+)-COO^-$	2.2	9.2	10.8
Histidine	His [H]	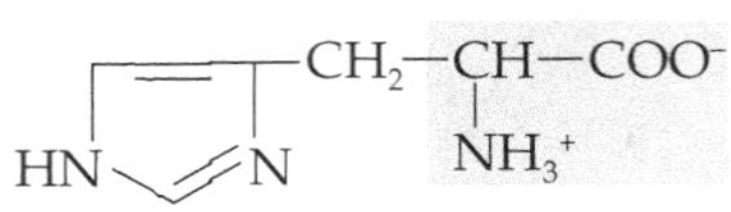	1.8	9.3	6.0
Containing Aromatic Rings					
Histidine	His [H]	↑ See above			
Phenylalanine	Phe [F]	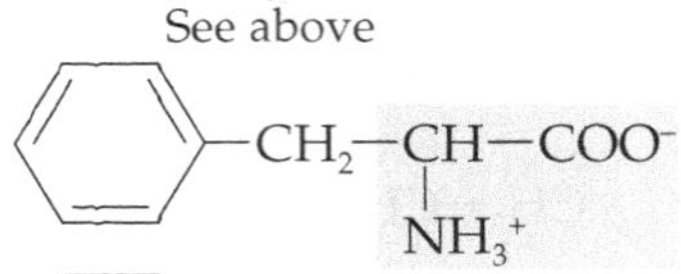	2.2	9.2	
Tyrosine	Tyr [Y]	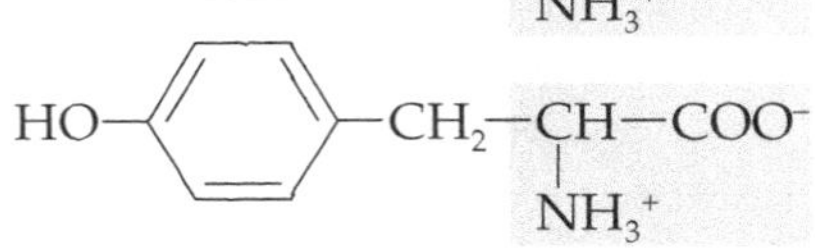	2.2	9.1	10.1

Typtophan	Trp [W]	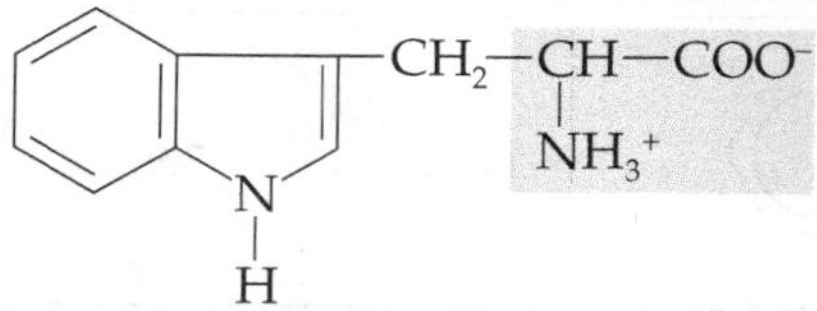	2.4	9.4	
Imino Acid Proline	Pro [P]	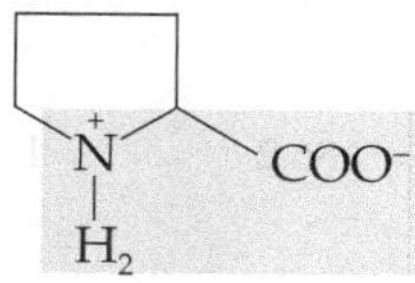	2.0	10.6	

Name Three-letter Symbol, and One-letter Symbol	Structural Formula[a]	Residue Mass (D)[b]	Average Occurrence in Proteins (%)[c]	pK_1 α-COOH[d]	pK_2 α-NH_3^+[d]	pK_R Side chain[d]
Amino acids with nonpolar side chains						
Glycine Gly G	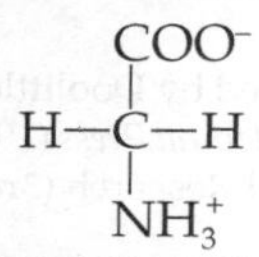	57.0	7.2	2.35	9.78	
Alanine Ala A	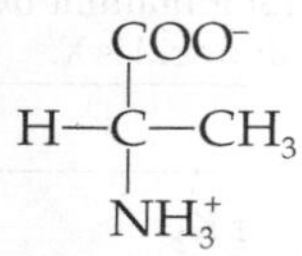	71.1	7.8	2.35	9.87	
Valine Val V	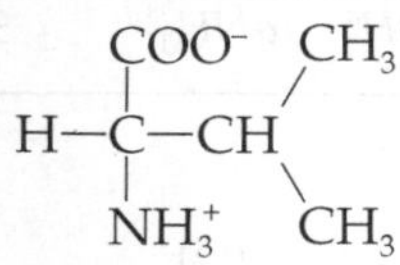	99.1	6.6	2.29	9.74	
Leucine Leu L	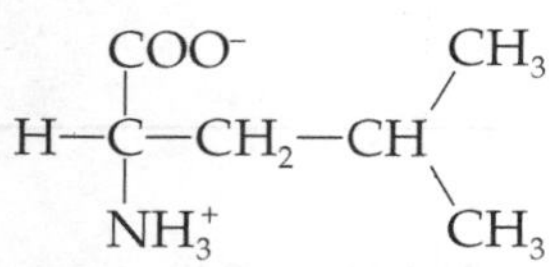	113.2	9.1	2.33	9.74	
Isolecucine Ile I	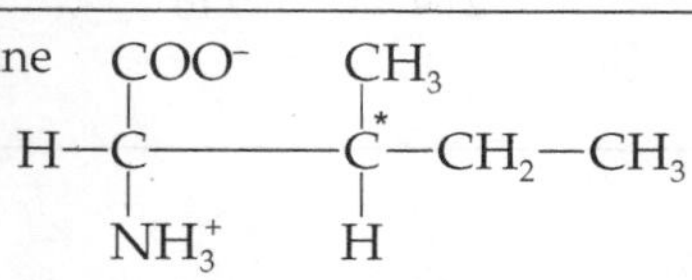	113.2	5.3	2.32	9.76	
Methionine Met M	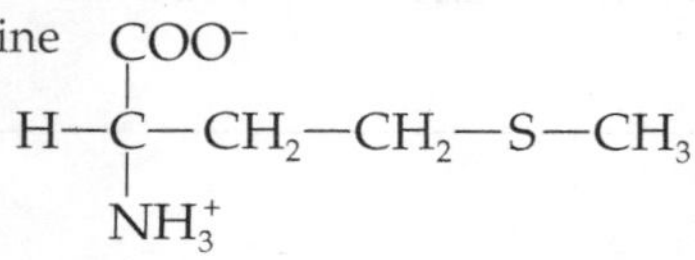	131.2	2.2	2.13	9.28	
Proline Pro P	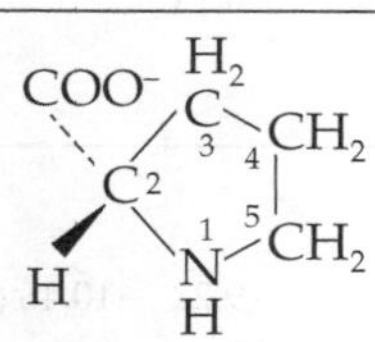	97.1	5.2	1.95	10.64	

Name, Three-letter Symbol, and One-letter Symbol	Structural Formula	Residue Mass (D)	Average Occurence in Proteins (%)	pK_1 α-COOH	pK_2 α-NH_3^+	pK_R Side chain
Phenylalanine Phe F	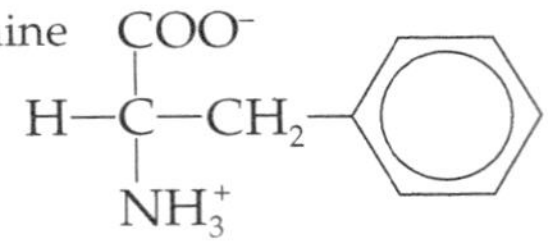	147.2	3.9	2.20	9.31	
Tryptophan Trp W	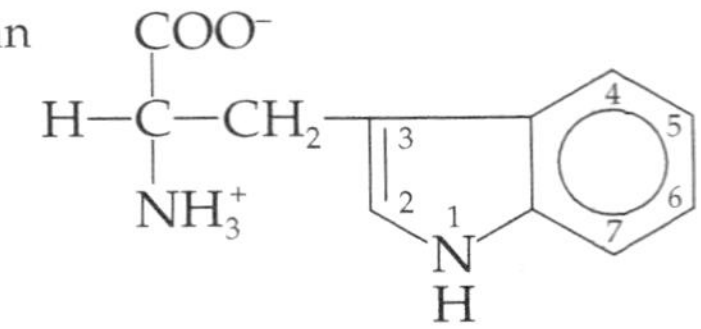	186.2	1.4	2.46	9.41	

(a) The ionic forms shown are those predominating at pH 7.0 although residue mass is given for the neutral compound. The C_α atoms, as well as those atoms marked with an asterisk, are chiral centers configurations as indicated according to Fischer projection formulas. The standard organic numbering system is provided for heterocycles.

(b) The residue masses are given for the neutral residues. For the molecular masses of the parent amino acids, add 18.0 D, the molecular mass of H_2O, to the residue masses. For side chain masses, subtract 56.0 D, the formula mass of a peptide group from the residue masses.

(c) Calculated from a database of nonredundant proteins containing 300,688 residues compiled by Doolittle, R. F. in Fasman, G,D. (Ed), *Predictions of protein Structure and the Principles of Protein Conformation, Plenum Press* (1989).

(d) *Source*: Dawson, R.M.C., Elliott, D.C., Elliott, W.H. And jones, K.M., Data for Biochemical Research (3rd ed.), pp 1–31, Oxford Science Publications (1986).

(e) The three-and one-letter symbols for asparagine or aspartic acid are Asx and B, whereas for glutamine *or* glutamic acid they are Glx and Z. The one-letter symbol for an undetermined or "nonstandard" amino acid is X.

Name, Three-letter Symbol, and One-letter Symbol	*Structural Formula*[a]	*Residue Mass (D)*[b]	*Average Occurence in Proteins (%)*[c]	pK_1 *α-COOH*[d]	pK_2 *α-NH_3^+*[d]	pK_R *Side chain*[d]
Amino acids with uncharge polar side chains						
Serine Ser S	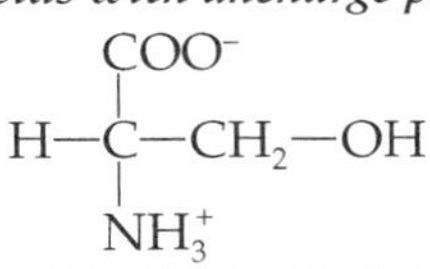	87.1	6.8	2.19	9.21	
Threonine Thr T	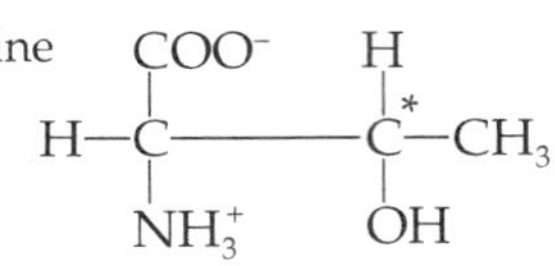	101.1	5.9	2.09	9.10	
Asparagine Asn N	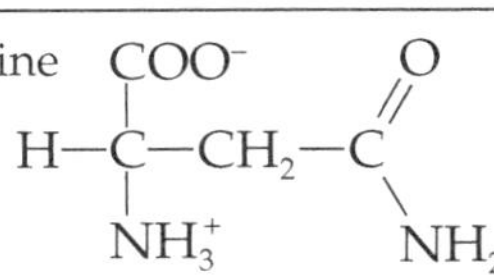	114.1	4.3	2.14	8.72	
Glutamine Gin Q	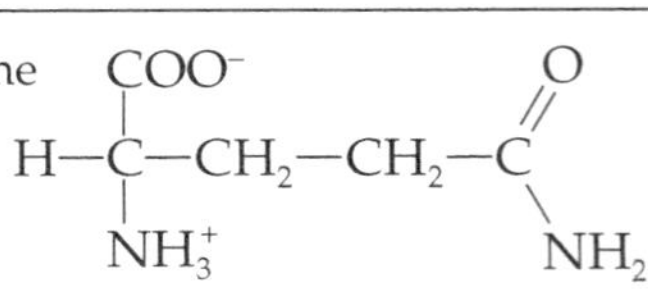	128.1	4.3	2.17	9.13	
Tyrosine Tyr Y	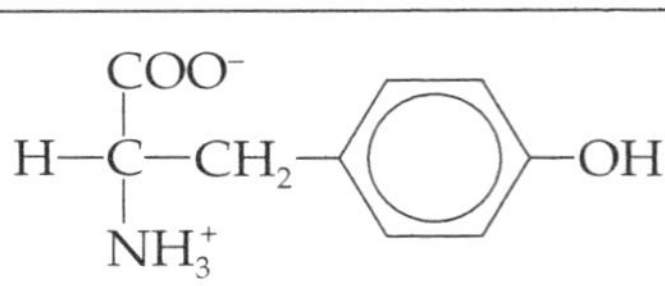	163.2	3.2	2.20	9.21	10.46 (phenol)

Cysteine Cys C	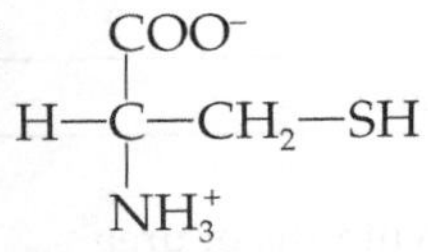	103.1	1.9	1.92	10.70	8.3 (sulfhydryl)
Amino acids with charge polar side chains						
Lysine Lys K	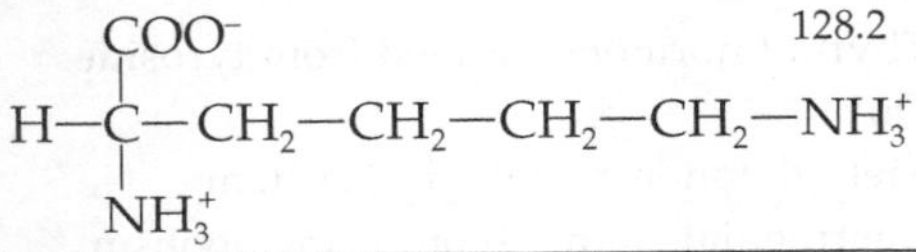	128.2	5.9	2.16	9.06	10.54 (ε-NH_3^+)
Arginine Arg A	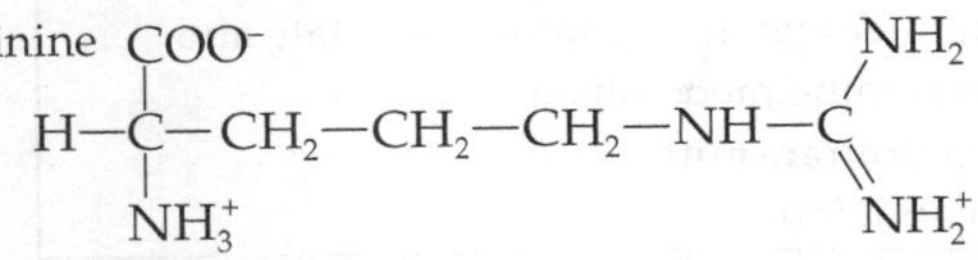	156.2	5.1	1.82	8.99	12.48 (guanidino)
Histidine His H	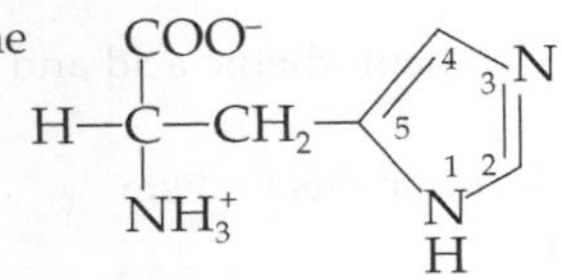	137.1	2.3	1.80	9.33	6.04 (imidazole)
Aspartic acid[e] Asp D	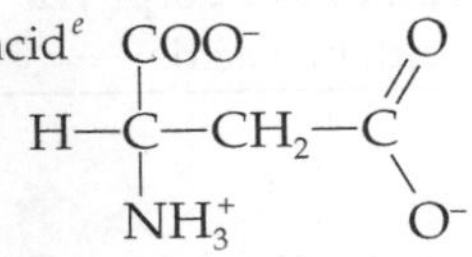	115.1	5.3	1.99	9.90	3.90 (β-COOH)
Glutamic acide Glu E	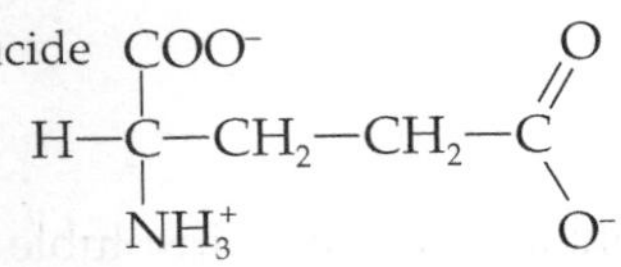	129.1	6.3	2.10	9.47	4.07 (γ-COOH)

Table 2.11

Essential Amino Acids	Nonessential Amino Acids
Arginine	Glycine
Valine	Alanine
Histidine	Serine
Isoleucine	Cysteine
Leucine	Aspartate
Lysine	Asparagines
Methionine	Glutamate
Phenylalnine	Glutamine
Threonine	Tyrosine
Tryptophan	Proline

Table 2.12 Non protein amino acids

Amino Acids	Function(s)
1. *α-amino acids*	
Ornithine	Intermediate in the biosynthesis of urea
Citrulline	Do
Arginosuccinic acid	Do
Thyroxine	Thyroid hormone derived from tyrosine
Triiodothyronine	Do
S-adenosylmethionine	Methyl donor in biological system
Homocysteine	Intermediate in methionine metabolism
Homoserine	Intermediate in threonine, aspartate and methionine metabolism
3,4-dihydroxy phynylalanine (DOPA)	A neurotransmitter
Azaserine	An antibiotic
2. *Non α-amino acids*	
β-alanine	Component of vitamin pentothenic acid and coenzyme-A
β-aminoisobutyric acid	End product of pyrimidine metabolism
γ-aminobutyric acid (GABA)	A neurotransmitter
δ-aminolevulinic acid (ALA)	Intermediate in the synthesis of porphyrin
Taurine	Found in association with bile acids

Properties of Amino Acids

The amino acids differ in their physico-chemical properties which eventually determine the characteristics of proteins.

Physical Properties

1. **Solubility:** Most of the amino acids are usually soluble in water and insoluble in organic solvents.
2. **Melting points:** Amino acids generally melt at higher temperatures, often above 200°C.
3. **Taste:** Amino acids may be sweet (Gly, Ala,Val), tasteless(Leu) or bitter (Arg, lle). Sodium glutamate is a salt of glutamic acid. It is employed as a flavouring agent in food industry to increase taste and flavour.
4. **Optical properties:** All the amino acids except glycine possess optical isomers due to the presence of asymmetric α-carbon atom. Some amino acids also have a second asymmetric carbon *e.g.*, isoleucine, threonine. The structure of L and D-amino acids in comparison with glyceraldehyde has been given (Fig. 2.23).
5. **Amino acids as ampholytes:** Amino acids contain both acidic (—COOH) and basic ($—NH_2$) groups. They can donate a proton or accept a proton; hence amino acids are regarded as ampholytes.

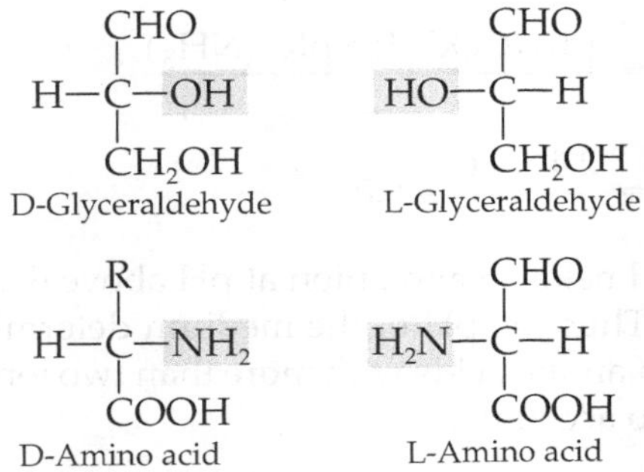

Fig. 2.23 Amino acid (D- and L-forms) based on the structure of glyceraldehyde

Zwitterion or Dipolar Ion

The name zwitter is derived from the German word which means *hybrid*. Zwitter ion (or dipolar ion) is a hybrid molecule containing positive and negative ionic groups. The amino acids rarely exist in a neutral form with free carboxylic ($-COOH$) and free amino ($-NH_2$) groups. In strongly acidic pH (low pH), the amino acid is positively charged (cation) while in strongly alkaline pH (high pH), it is negatively charged (anion). Each amino acid has a characteristic pH (*e.g.*, leucine, pH 6.0) at which it carries both positive and negative charges and exists as zwitterion (Fig. 2.24).

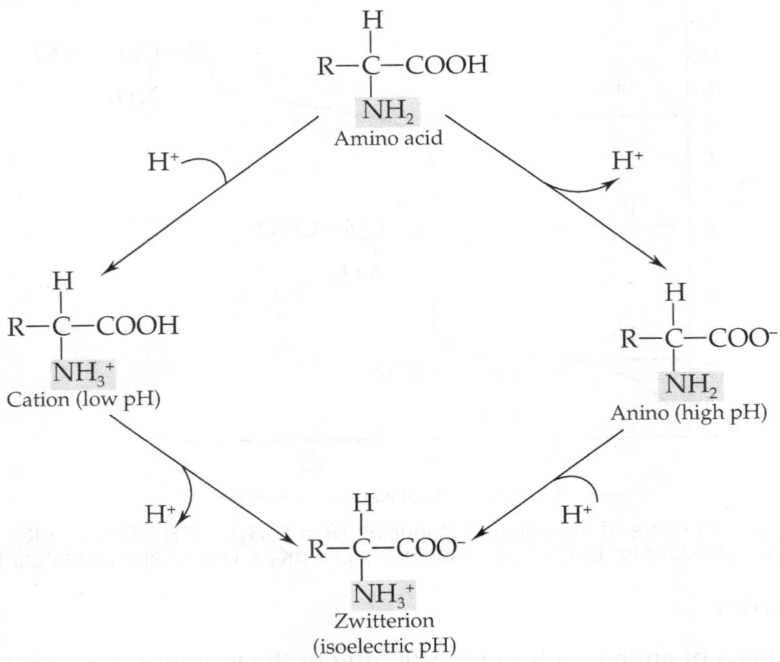

Fig. 2.24 Existence of an amino acid as cation, anion and zwitterion

Isoelectric pH

Isoelectric pH (symbol pI) is defined as the pH at which a molecule exists as a zwitterion or dipolar ion and carries no net charge. Thus, the molecule is electrically neutral.

The pI value can be calculated by taking the average pKa values corresponding to the ionisable groups. For instance, leucine has two ionisable groups and its pI can be calculated as follows:

$$pH = \frac{pK_1(COOH) + pK_2(NH_3)}{2}$$

$$pI = \frac{2.4 + 9.6}{2} = 6.0$$

Leucine exists as cation at pH below 6 and anion at pH above 6. At the Isoelectric pH (pI = 6.0), leucine is found as zwitterion. Thus the pH of the medium determines the ionic nature of amino acids. For the calculation of pI of amino acids with more than two ionisable groups, the pKa's for all the groups have to be taken into account.

Titration of Amino Acids

The existence of different ionic forms of amino acids can be more easily understood by the titration curves. The graphic representation of leucine titration is depicted in Fig. 2.25. At low pH, leucine exists in a fully protonated form as cation. As the titration proceeds with NaOH, leucine loses its protons and at isoelectric pH (pI), it becomes a zwitterion. Further titration results in the formation of anionic form of leucine.

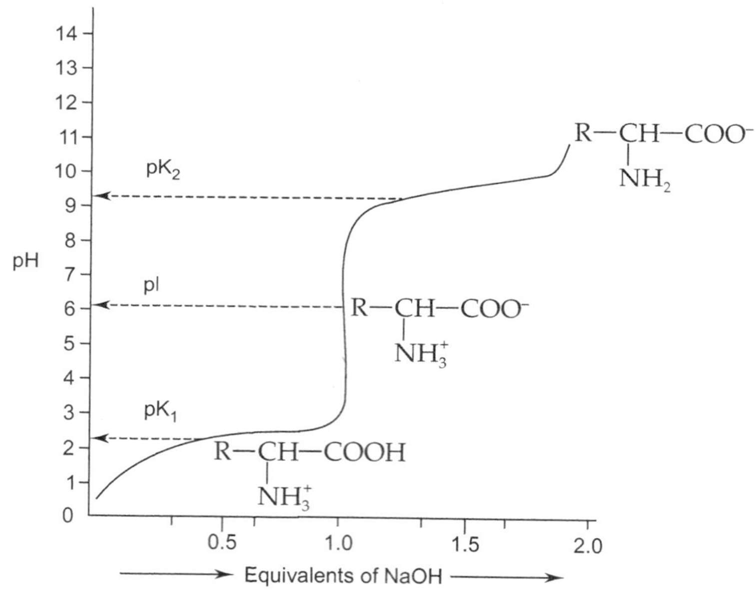

Fig. 2.25 Titration curve of an amino acid-leucine (R = $(CH_3)_2$—CH—CH_2—; pK_1 = Dissociation constant for COOH; pI = Isoelectric pH; pK_2 = Dissociation constant for NH_3^+)

Chemical Properties

The general reactions of amino acids are mostly due to the presence of two functional groups namely carboxyl (—COOH) group and amino (—NH_2) group.

Reactions due to —COOH group

1. Amino acids form salts (—COONa) with bases and esters (—COOR') with alcohols.
2. ***Decarboxylation:*** Amino acids undergo decarboxylation to produce corresponding amines.

$$\underset{\displaystyle NH_3^+}{R-\overset{}{CH}-COO^-} \rightarrow \underset{\displaystyle NH_3^+}{R-CH_2} + CO_2$$

This reaction assumes significance in the living cells due to the formation of many biologically important amines. These include histamine, tyramine and γ-amino butyric acid (GABA) from the amino acids histidine, tyrosine and glutamate respectively.

3. ***Reaction with ammonia:*** The carboxyl group of dicarboxylic amino acids reacts with NH_3 to form amide.

$$\text{Aspartic acid} + NH_3 \rightarrow \text{Asparagine}$$
$$\text{Glutamic acid} + NH_3 \rightarrow \text{Glutamine}$$

Reactions due to -NH_2 group

4. The amino groups behave as base and combine with acids (*e.g.*, HCL) to form salts ($-NH_3^+CL^-$).
5. ***Reaction with ninhydrin:*** The α-amino acids react with ninhydrin to form a purple, blue or pink colour complex (Ruhemann's purple).

$$\text{Amino acids} + \text{Ninhydrin} \rightarrow \text{Keto acid} + NH_3 + CO_2 + \text{Hydrindantin}$$
$$\text{Hydrindantin} + NH_3 + \text{Ninhydrin} \rightarrow \text{Ruhemann's purple}$$

Ninhydrin reaction is effectively used for the quantitative determination of amino acids and proteins. (Note: proline and hydroxyproline give yellow colour with ninhydrin).

6. ***Colour reactions for amino acids:*** Amino acid can be identified by specific colour reactions Table 2.13.
7. ***Transamination:*** Transfer of an amino group from an amino acid to a keto acid to form a new amino acid is a very important reaction in amino acid metabolism.
8. ***Oxidative deamination:*** The amino acids undergo oxidative deamination to liberate free ammonia.

Table 2.13 Colour reactions of amino acids/proteins.

S.No.	Reaction	Specific group or Amino Acid
1.	Sulpher test	Sulfhydryl groups (Cys)
2.	Folin-Coicalteau's test	Phenolic group (Tyr)
3.	Pauly's test	Imidazole ring (His)
4.	Milons reaction	Phenolic group (Tyr)
5.	Hopkins-Cole reaction	Indole ring(Trp)
6.	Xanthoproteic reaction	Benzene ring of aromatic Amino acids (Phe, Tyr,Trp)
7.	Nitroprusside reaction	Sulfhydryl groups (Cys)
8.	Biureret action	Two peptide linkages
9.	Ninhydrin reaction	A-amino acid
10.	Sakaguchi reaction	Guanidino group (Arg)

Structure of Proteins

Proteins have a very spacific structure, it is the polymers of L-α-amino acids. To understand the functions of proteins at a molecular level, it is often necessary to determine their three-dimensional structure. This is the subject of the scientific field of structural biology, which employs techniques such as X-ray crystallography, NMR spectroscopy, and dual polarisation interferometry to determine the structure of proteins. The structure of proteins is rather complex which can be divided four levels of organization:

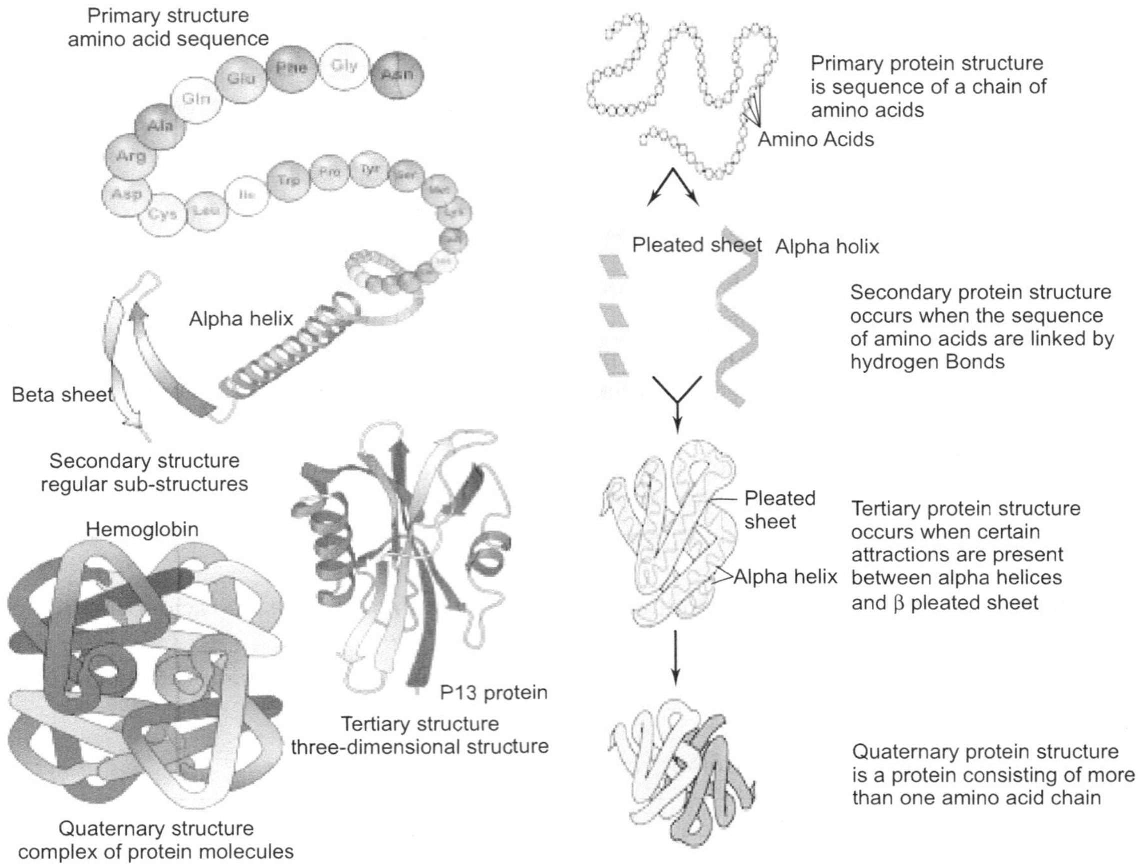

Fig. 2.26 The four levels of Protein structure

Primary Structure

This is the chemical sequence of different amino acids, is called the primary structure of the peptide or protein. The primary structure of a protein is determined by the gene corresponding to the protein. A specific sequence of nucleotides in DNA is transcribed into mRNA, which is read by the ribosome in a process called translation. The sequence of a protein is unique to that protein, and defines the structure and function of the protein. The Edman reaction enabled amino acid sequence analysis to be automated. Mass spectrometry provides a sensitive a versatile tool for determining primary structure and for the identification of post translational modifications. Two amino acids can be combined in a condensation reaction. By repeating this reaction, long chains of residues (amino acids in a peptide bond) can be generated Fig. 2.27. This reaction is catalysed by the ribosome in a process known as translation.

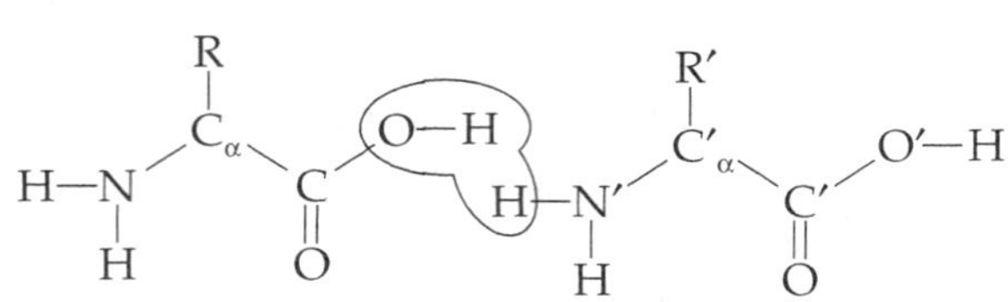

Fig. 2.27 Formation of peptide bond

Secondary Structure

Secondary structure of protein has come to mean the specification of regular polypeptide backbone folding patterns: helices, plated sheets, and turns. Peptide bonds are folded in a regular manner in colloid structure of globular proteins. The foldings are the results of linking of the carboxyl and amino groups of the peptide chains by means of hydrogen bonds and disulfide bonds. Such foldings are referred to as the secondary structure of the protein. The DSSP definition (The Dictionary of Protein Secondary Structure, in short DSSP, is commonly used to describe the protein secondary structure with single letter codes. The secondary structure is assigned based on hydrogen bonding patterns as those initially proposed by Pauling et al. in 1951) of hydrogen bond is used in the secondary structure of protein because it is described by patterns of hydrogen bonds between backbone amide and carboxyl groups. Residues in protein helices generally adopt backbone dihedral angles in a particular region of the Ramachandran plot (Fig. 2.29); therefore, a segment of residues with hydrogen bonding, such dihedral angles produces a regular coiled arrangement, called α-helix.

DSSP H-bond definition

Hydrogen bonding plays a very crucial role in the formation of secondary structure of protein, thus the accurate definition of a hydrogen bond is critical. The standard H-bond definition for secondary structure is that of DSSP, which is a purely electrostatic model. It assigns charges of $\pm q_1 \equiv 0.42e$ to the carbonyl carbon and oxygen, respectively, and charges of $\pm q_{12} \equiv 0.20e$ to the amide nitrogen and hydrogen, respectively. The electrostatic energy is

$$E = q_1 q_2 \left[\frac{1}{r_{ON}} + \frac{1}{r_{CH}} - \frac{1}{r_{OH}} - \frac{1}{r_{CN}} \right] \cdot 332 \text{ kcal/mol.}$$

According to DSSP, an H-bond exists if E is less than –0.5 kcal/mol. Although, the DSSP formula is a relatively crude approximation of the physical H-bond energy, it is generally accepted as a tool for defining secondary structure.

Ramachandran Plot

Ramachandran plot is invented by Gopalasamudram Narayana Ramachandran, by this approach to visualize dihedral angles ψ against φ of amino acid residues in protein structure. This is also known as a the name of Ramachandran map or a Ramachandran diagram or a [φ, ψ] plot. It shows the possible conformations of ψ and φ angles for a polypeptide (Fig. 2.28) and the values of these angles can be determine by calculating the distances between the atoms of a tripeptide at all values of ψ and φ for the central peptide units.

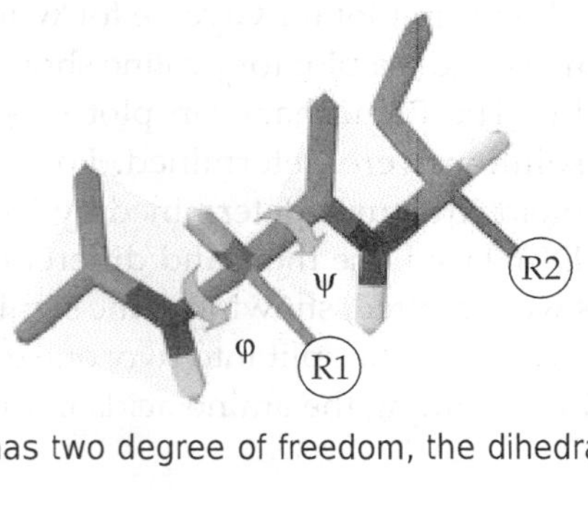

Fig. 2.28 A peptide bond has two degree of freedom, the dihedral angles named φ and ψ by Ramachandran

Figure 2.29 indicates that approximately 75% of the Ramchandran plot is conformationally inaccessible to a polypeptide chain. The particular regions of the Ramchandran plot that represents possible confirmation depend on the van der Waals radii chosen to calculated. As well as we see that most of the common types of regular secondary structure found in protein fall within allowed regions of the Ramchandran plot.

Mathematically, the Ramachandran plot is the visualization of a function $f: (-\pi, \pi) \times (-\pi, \pi) \rightarrow R_+$. The domain of this function is the torus. Hence, the conventional Ramachandran plot is a projection of the torus on the plane, resulting in a distorted view and the presence of discontinuities.

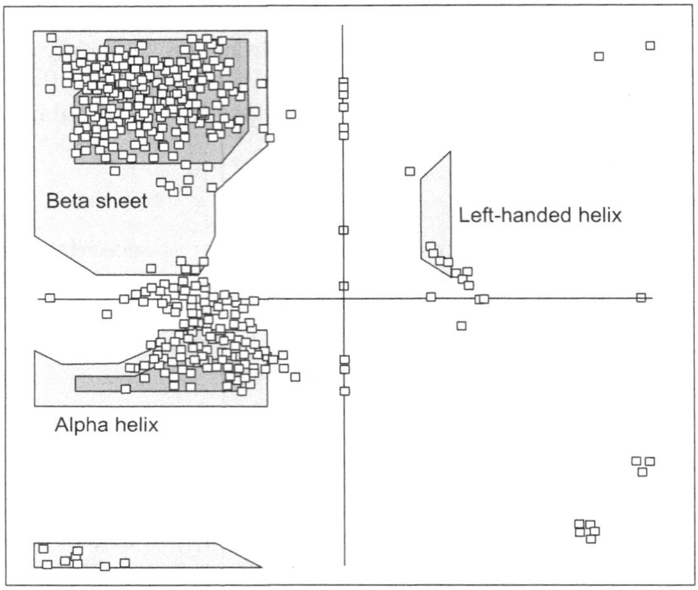

Fig. 2.29 A Ramachandran plot generated from the protein PCNA, a human DNA clamp protein that is composed of both beta sheets and alpha helices (PDB ID 1AXC). Points that lie on the axes indicate N- and C-terminal residues for each subunit. The green regions show possible angle formations that include Glycine, while the blue areas are for formations that don't include Glycine

One would expect that larger side chains would result in more restrictions and consequently a smaller allowable region in the Ramachandran plot. In practice this does not appear to be the case; only the methylene group at the β position has an influence. Glycine has a hydrogen atom, with a smaller van der Waals radius, instead of a methyl group at the β position. Hence it is least restricted and this is apparent in the Ramachandran plot for Glycine for which the allowable area is considerably larger. In contrast, the Ramachandran plot for proline shows only a very limited number of possible combinations of ψ and φ. The Ramachandran plot was calculated just before the first protein structures at atomic resolution were determined. Forty years later there were tens of thousands of high-resolution protein structures determined by X-ray crystallography and deposited in the Protein Data Bank (PDB). From one thousand different protein chains, Ramachandran plots of over 200 000 amino acids were plotted, showing some significant differences, especially for glycine. The upper left region was found to be split into two; one to the left containing amino acids in beta sheets and one to the right containing the amino acids in random coil of this conformation.

Alpha Sheet

The simplest arrangement of polypeptide chain could assume with its rigid peptide bonds (but other single bonds free to rotate) are a helical structure, which Linus Pauling and Robert Corey in 1951called the α-helix. The alpha sheet also known as an alpha pleated sheet or a polar pleated sheet. In the structure of α-helix, the polypeptide back bone is tightly bound around an imaginary axis drawn longitudinally through the middle of the helix, and the R-groups of the amino acid residues protrude out ward from the helical back bone. There are 3.6 amino acid residues per turn of the helix found. The distance travelled per turn is 0.54 nm. The space found between per amino acid residues in 0.15 nm. About one fourth of all amino acid residues in polypeptides are found in α-helices. The structure of α-helix is stabilised by hydrogen bon between the hydrogen atom attached to the electronegative nitrogen atom of a peptide linkage and the electronegative carbonyl oxygen atom of the fourth amino acid on the amino terminal side of that peptide bond. Almost in the structure of α-helix every peptide bon participates in such type of hydrogen bonding. Naturally occurring L-amino acids can form either right or left handed α-helices. Although, the alpha sheet is only rarely observed in natural protein structures, it has been speculated to play a role in amyloid disease and it was found to be a stable form for amyloidogenic proteins in one set of molecular dynamics simulations. Alpha sheets have also been observed in X-ray crystallography structures of designed peptides.

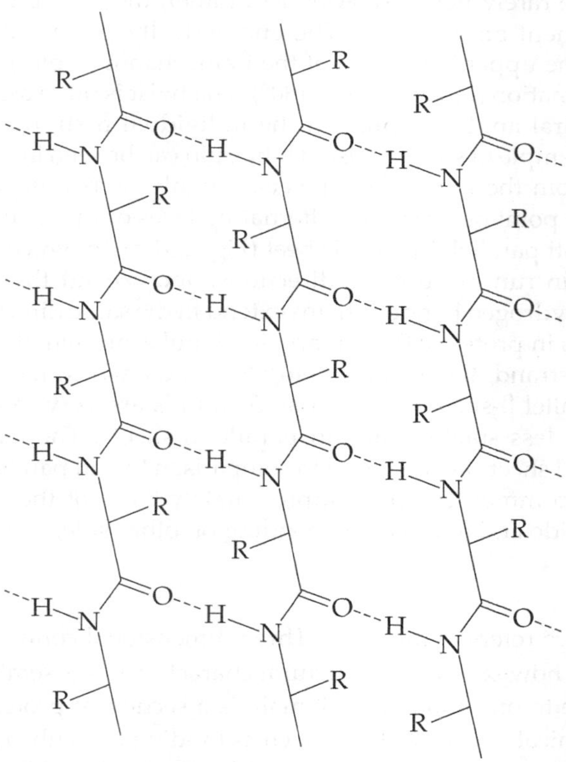

Fig. 2.30 Diagram of the hydrogen bonding patterns in the alpha sheet structure. Oxygen atoms are shown in red and nitrogen in blue; dotted lines represent hydrogen bonds. R groups represent the amino acid side chains

Beta Sheet (Structure and Orientation)

In 1951, Linus Pauling and Robert Corey postulated the existence of a second type of repetitive structure that is the β-pleated sheet. In β-pleated sheet, hydrogen bonding occurs between neighbouring polypeptide chain. A beta strand (also β-strand) is a stretch of amino acids typically 5–10 amino acids long whose peptide backbones are almost fully extended. The association of beta sheets has been implicated in the formation of protein aggregates and fibrils observed in many human diseases, notably the amyloidoses. The majority of β strands are arranged adjacent to other strands and form an extensive hydrogen bond network with their neighbours in which the N-H groups in the backbone of one strand establish hydrogen bonds with the C=O groups in the backbone of the adjacent strands. In the fully extended β strand, successive side chains point straight up, then straight down, then straight up, etc. Adjacent β strands in a β-sheet are aligned so that their Cα atoms are adjacent and their side chains point in the same direction. The "pleated" appearance of β strands arises from tetrahedral chemical bonding at the Cα atom; for example, if a side chain points straight up, then the bond to the C′ must point slightly downwards, since its bond angle is approximately 109.5°. The pleating causes the distance between C_i^{α} and C_{i+2}^{α} to be approximately 6 Å, rather than the 7.6 Å (2 × 3.8 Å) expected from two fully extended trans peptide virtual bonds. The "sideways" distance between adjacent Cα atoms in hydrogen-bonded β strands is roughly 5 Å.

However, β strands are rarely perfectly extended; rather, they exhibit a slight twist due to the chirality of their component amino acids. The energetically preferred dihedral angles (φ, ψ) = (–135°, 135°) (broadly, the upper left region of the Ramachandran plot) diverge somewhat from the fully extended conformation (φ, ψ) = (–180°, 180°). The twist is often associated with alternating fluctuations in the dihedral angles to prevent the individual β strands in a larger sheet from splaying apart. A good example of such a twisted β-hairpin can be seen in the protein BPTI. The side chains point outwards from the folds of the pleats, roughly perpendicularly to the plane of the sheet; successive residues point outwards on alternating faces of the sheet. There are two types of β-pleated sheet occurs; anti parallel β-pleated sheet (Fig. 2.31 a), in which neighbouring hydrogen bonded polypeptide chain run in opposite directions, and second the parallel β-pleated sheet (Fig. 2.31 b) in which the hydrogen bonded chains extend in the same direction. β-pleated sheets are common structural motifs in proteins. Particularly in globular protein, they consist of from 2 to as many as 15 polypeptide strand, the average being 6 strands, which have an aggregate width of approximately 25 Å. Parallel β-sheets of less than 5 strands are very rare. This finding suggests that parallel β-sheets are less stable than anti parallel β-sheets. The reason behind this is, the hydrogen bonds of parallel sheets are distorted in comparison to anti parallel sheets. Mixed parallel-anti paralle β-sheets are common but only approximately 20 % of the strands in β-sheets have parallel bonding on one side and anti parallel bonding on other side.

Tertiary Structure

The word tertiary structure refers to the entire Three-dimensional confirmation of a polypeptide. It postulate, in 3-D space, how secondary structural characteristics assemble to form domains and how these domains relate to one another. A domain is a section of protein structure sufficient to perform a particular chemical or physical task such as binding of a substrate or other ligands. The formation of tertiary structure is generally focused by the burial of hydrophobic residues, but other interactions such as hydrogen bonding, ionic interactions and disulfide bonds can also stabilize the tertiary structure. The tertiary structure encompasses all the noncovalent interactions that are not considered secondary structure, and is what defines the overall fold of the protein, and is usually indispensable for the function of the protein.

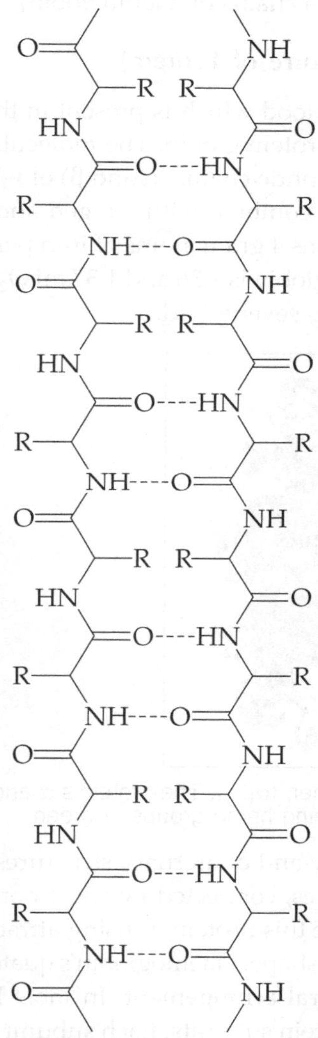

Fig. 2.31 (*a*) of the hydrogen bonding patterns, represented by dotted lines, in an antiparallel beta sheet. Oxygen atoms are colored red and nitrogen atoms colored blue

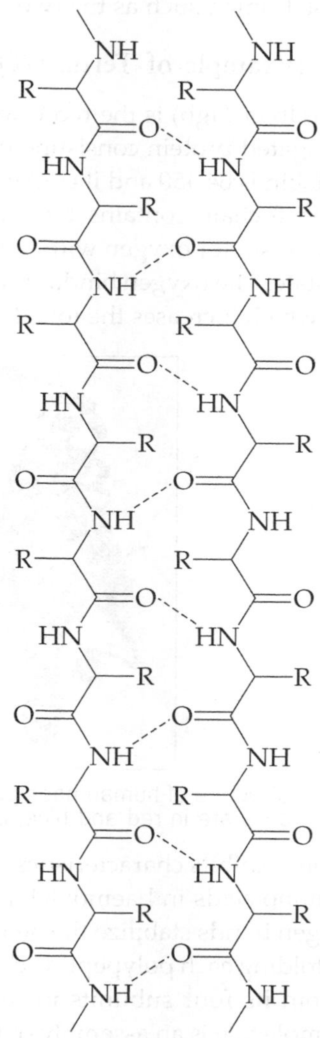

Fig. 2.31 (*b*) of the hydrogen bonding patterns, represented by dotted lines, in a parallel beta sheet. Oxygen atoms are colored red and nitrogen atoms colored blue

Quaternary Structure

The quaternary structure defines the polypeptide composition of a protein and, for an oligomeric protein, the spatial relationship between its subunit or protomers. Monomeric protein consists of a single polypeptide chain. Dimeric protein contains two polypeptide chains. Homodimers contain two copies of the same polypeptide chain, while in a heterodimer the polypeptide chain are differ.

The subunits are usually related to one another by symmetry axes, such as a 2-fold axis in a dimer. Multimers made up of identical subunits may be referred to with a prefix of "homo-" (*e.g.*, a homotetramer) and those made up of different subunits may be referred to with a prefix of "hetero-" (*e.g.*, a heterotetramer, such as the two alpha and two beta chains of haemoglobin).

Haemoglobin (Example of Tertiary/Quaternary Structure of Protein)

Haemoglobin (Hb or Hgb) is the red colouring matter of blood which is present in the red blood cells. It is conjugated protein consisting of heme and the protein globin. The molecular weight of adult haemoglobin is 64,450 and it contains two pairs of peptide chains (α and β) of which α chain contain 141 and β chain contains 146 amino acids. It can combine with oxygen and acts as the transport mechanism for oxygen within the blood. It contains 4 gram atoms of iron per mole in the ferrous (Fe^{++}) state. The oxygen binding capacity of haemoglobin is 1.36 and 1.37 ml O_2 per gram of haemoglobin, which increases the total *blood oxygen capacity* seventyfold.

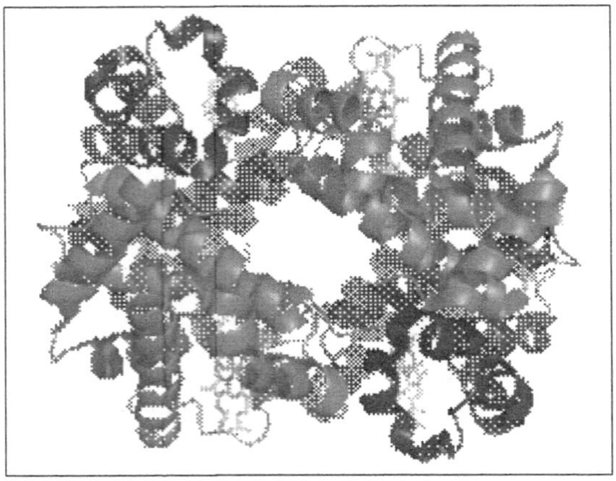

Fig. 2.32 Structure of human haemoglobin (heterotetramer, $(\alpha\beta)_2$). The protein's α and β subunits are in red and blue, and the iron-containing heme groups in green

Haemoglobin exhibits characteristics of both the tertiary and quaternary structures of proteins. Most of the amino acids in haemoglobin form alpha helices, connected by short non-helical segments. Hydrogen bonds stabilize the helical sections inside this protein, causing attractions within the molecule, folding each polypeptide chain into a specific shape. Haemoglobin's quaternary structure comes from its four subunits in roughly a tetrahedral arrangement. In most humans, the haemoglobin molecule is an assembly of four globular protein subunits. Each subunit is composed of a protein chain tightly associated with a non-protein heme group. Each protein chain arranges into a set of alpha-helix structural segments connected together in a globin fold arrangement, so called because this arrangement is the same folding motif used in other heme/globin proteins such as myoglobin. This folding pattern contains a pocket that strongly binds the heme group. A heme group consists of an iron (Fe) ion (charged atom) held in a heterocyclic ring, known as a porphyrin. A porphyrin ring consists of four pyrrole molecules cyclically linked together with the iron ion bound in the centre. The iron ion, which is the site of oxygen binding (*i.e.*, transporting the oxygen and CO_2 in the blood), coordinates with the four nitrogens in the centre of the ring, which all lie in one plane. The iron is bound strongly to the globular protein via the imidazole ring of the F8 histidine residue below the porphyrin ring. A sixth position can reversibly bind oxygen by a coordinate covalent bond, completing the octahedral group of six ligands. Oxygen binds in an "end-on bent" geometry where one oxygen atom binds Fe and the other protrudes at an angle. When oxygen is not bound, a very weakly bonded water molecule fills the site, forming a distorted octahedron. The iron ion may either be in the Fe^{2+} or Fe^{3+} state, but ferrihemoglobin (methemoglobin) (Fe^{3+}) cannot bind oxygen. In binding, oxygen temporarily oxidizes (Fe^{2+}) to (Fe^{3+}), so iron must exist in the +2 oxidation state to bind oxygen. The enzyme methemoglobin reductase reactivates haemoglobin found in the inactive (Fe^{3+}) state by reducing the iron centre.

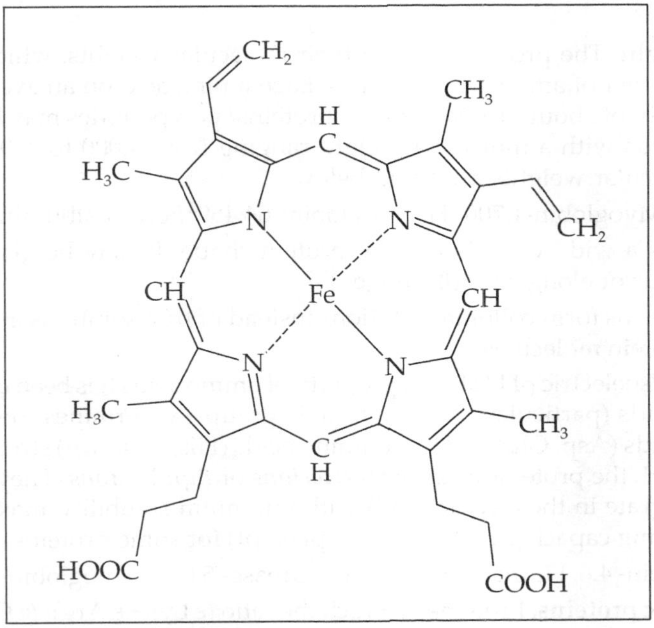

Fig. 2.33 Structure of Heme group

In adult humans, the most common haemoglobin type is a tetramer (which contains 4 subunit proteins) called haemoglobin A, consisting of two α and two β subunits non-covalently bound, each made of 141 and 146 amino acid residues, respectively. This is denoted as $\alpha_2\beta_2$. The subunits are structurally similar and about the same size. Each subunit has a molecular weight of about 17,000 daltons, for a total molecular weight of the tetramer of about 68,000 daltons (64,458 g/mol) thus, 1 g/dL = 0.1551 mmol/L. Haemoglobin A is the most intensively studied of the haemoglobin molecules. As an infant, the haemoglobin molecule is made up of 2 α chains and 2 gamma chains. The gamma chains are gradually replaced by β chains as the infant grows. The four polypeptide chains are bound to each other by salt bridges, hydrogen bonds, and hydrophobic interactions. There are two kinds of contacts between the α and β chains: $\alpha_1\beta_1$ and $\alpha_1\beta_2$. Generally speaking, haemoglobin can be saturated with oxygen molecules (oxyhemoglobin), or desaturated with oxygen molecules (deoxyhemoglobin). Oxyhemoglobin is formed during physiological respiration when oxygen binds to the heme component of the protein haemoglobin in red blood cells. This process occurs in the pulmonary capillaries adjacent to the alveoli of the lungs. The oxygen then travels through the blood stream to be dropped off at cells where it is utilized in aerobic glycolysis and in the production of ATP by the process of oxidative phosphorylation. It does not, however, help to counteract a decrease in blood pH. Ventilation, or breathing, may reverse this condition by removal of carbon dioxide, thus causing a shift up in pH. Deoxyhemoglobin is the form of haemoglobin without the bound oxygen. The absorption spectra of oxyhemoglobin and deoxyhemoglobin differ. The oxyhemoglobin has significantly lower absorption of the 660 nm wavelength than deoxyhemoglobin, while at 940 nm its absorption is slightly higher. This accounts for haemoglobin's red colour and deoxyhemoglobin's blue colour. This difference is used for measurement of the amount of oxygen in patient's blood by an instrument called pulse oximeter. Haemoglobin provides highlights in to the molecular basis of genetic disease such as sickle cell disease and Thalassemia. Cyanide and carbon monoxide disrupt the physiologic functions of the heme, proteins, cytochrome oxidase, and haemoglobin respectively.

Properties of Proteins

1. **Molecular weight:** The proteins vary in their molecular weights, which, in turn, is dependent on the number of amino acid residues. Each amino acid on an average contributes to a molecular weight of about 110. Majority of proteins/ polypeptides may be composed of 40 to 4,000 amino acids with a molecular weight ranging from 4,000 to 440,000. A few proteins with their molecular weights are listed below:

   Insulin- 5,700; Myoglobin-1,700; Haemoglobin- 64,450; Serum albumin- 69,000.
2. **Shape:** There is a wide variation in the protein shape. It may be globular (insulin), oval (albumin) fibrous or elongated (fibrinogen).
3. **Solubility:** Proteins form colloidal solutions instead of true solutions in water. This is due to huge size of protein molecules.
4. **Isoelectric pH:** Isoelectric pH (pI) as a property of amino acids has been described. The nature of the amino acids (particularly their ionisable groups) determines the pI of a protein. The acidic amino acids (Asp, Glu) and basic amino acids (His, Lys, Arg) strongly influence the pI. At isoelectric pH, the proteins exist as ***zwitterions*** or ***dipolar ions***. They are electrically neutral (do not migrate in the electric field) with minimum solubility, maximum perceptibility and least buffering capacity. The isoelectric pHs (pI) for some proteins are given here:

   Pepsin-1.1; Casein- 4.6; Human albumin- 4.7; Urease- 5.0; Haemoglobin- 6.7; Lysozyme- 11.0.
5. **Acidic and basic proteins:** Proteins in which the ratio (ε Lys + ε Arg)/ (ε Glu + ε Asp) is greater than 1 are referred to as basic proteins. For acidic proteins, the ratio is less than 1. The acidic and basic properties of proteins are utilized for staining them. The acidic dye eosin (H^+— dye^-) combines with cationic form ($-NH_3^+$) of protein while the basic dye methylene blue (dye^+—OH^-) combines with anionic form ($-COO^-$) proteins.
6. **Colour reactions of proteins:** The proteins give a number of colour reactions which are often useful to recognize the nature of the amino acids present in them. The colour reactions given by proteins due to the presence of specific amino acids are given in Table 2.13. These reactions are often useful to know the presence or absence of the amino acids in the protein.
7. **Precipitation of proteins:** Proteins exist in colloidal solution due to hydration of polar groups ($-COO^-$, $-NH_3^+$, $-OH$). Proteins can be precipitated by dehydration or neutralization of polar groups. Various methods are in use to achieve protein precipitation.

   (*a*) ***Precipitation by salting out:*** The process of protein precipitation by the additional of neutral salts such as ***ammonium sulfate*** or ***sodium sulfate*** is known as salting out. This phenomenon is explained on the basis of dehydration of protein molecules by salts. This causes increased protein-protein interaction, resulting in molecular aggregation and precipitation. The amount of salt required for protein precipitation depends on the size (molecular weight) of the protein molecule. In general, the higher is the protein molecular weight, the lower is the salt required for precipitation. Thus, serum globulins are precipitated by half saturation with ammonium sulfate while albumin is precipitated by full saturation. Salting out procedure is conveniently used for separating serum albumins from globulins. The addition of small quantities of neutral salts increases the solubility of proteins. This process called as ***salting in*** is due to the diminished protein-protein interaction at low salt concentration.

(*b*) ***Precipitation by salts of heavy metals:*** Heavy metal ions like Pb^{2+}, Hg^{2+}, Fe^{2+}, Zn^{2+}, Cd^{2+} cause precipitation of proteins. These metals being positively charged, when added to protein solution (negatively charged) in alkaline medium results in precipitate formation.

(*c*) ***Precipitation at pI:*** The proteins in general are least soluble at isoelectric pH. Certain proteins (*e.g.*, casein) get easily precipitated when the pH is adjusted to pI (4.6 for casein). Formation of curd from milk is a marvellous example of slow precipitation of milk protein, casein at pI. This occurs due to the lactic acid produced by fermentation of bacteria which lowers the pH to the pI of casein.

(*d*) ***Precipitation by organic solvents:*** Organic solvents such as alcohol are good protein precipitating agents. They dehydrate the protein molecule by removing the water envelope and cause precipitation.

(*e*) ***Precipitation by anionic or alkaloid reagents:*** Proteins can be precipitated by trichloroacetic acid, sulphosalicylic acid, phosphotungstic acid, picric acid, tannic acid, phosphomolybdic acid etc. By the addition of these acids, the proteins existing as cation are precipitated by the anionic form of acids to produce protein-sulphosalicylate, protein-tungstate, protein-picrate etc.

Biological Functions of Proteins

Proteins participate multiplicity of roles in almost all biological process. Their importance and the scope of their activity are exemplified in the following functions.

1. ***Fundamental constituents:*** Proteins are the fundamental constituents of all protoplasm and are involved in the structure of the living cell and its function.
2. ***Homeostatic control:*** Proteins execute their activities in the homeostatic control of the volume of the circulating blood and that of the interstitial fluids through the plasma proteins.
3. ***Coordinated motion:*** Proteins are the major component of muscle. Muscle contraction is an accomplished by the sliding motion of two kinds of protein filaments actin and myosin.
4. ***Mechanical support:*** the high tensile strength of skin and bone is due to the presence of collagen, a fibrous protein.
5. ***Transport and storage:*** many small molecules and ions are transported by specific protein. For example, haemoglobin transports oxygen in erythrocytes, where as myoglobin, transport oxygen in muscle. Iron is carried in the plasma of blood by transferrin and is stored in the liver as a complex with ferritin.
6. ***Immune protection:*** Antibodies are highly specific proteins that recognize and combine with foreign substances like viruses, bacteria and cells from other organism.
7. ***Control of growth and differentiation:*** control sequential expression of genetic information is essential for the growth and differentiation of the cells. In bacteria, repressor proteins are important control elements that silence specific segments of the DNA of a cell. In higher organisms, growth and differentiation are controlled by growth factor protein. For example, nerve growth factor guides the formation of neural network. Indeed, proteins serve in all cells as sensors that control the flow of energy and matter.
8. ***Enzymatic catalysis:*** merely all chemical reaction in biological systems are catalyzed by specific macromolecules called enzyme. Enzyme exhibit enormous catalytic power. They usually increase reaction rate by at least a million fold. Several thousand enzymes have been characterized, and many of them have been crystallized.

9. ***Generation and transmission of nerve impulse:*** the response of nerve cells to specific stimuli is mediated by receptor proteins. For example, rhodopsin is a photo receptor protein in retinal rod cells.
10. ***Blood clotting***: Proteins are involved in blood clotting through thrombin, fibrinogen and other factors.

2.2.3 Lipids

Lipids (Greek: lipos, fat) are substances of biological origin that are soluble in organic solvents such as chloroform and methanol but are only sparingly soluble, if at all, in water. Lipids are a heterogeneous group of compounds, including fats, waxes, sterols, fat-soluble vitamins (such as vitamins A, D, E and K), monoglycerides, diglycerides, phospholipids, and others.

Classification of Lipids (Diagrammatic Compendium)

Bloor has proposed the following classification of lipids:

1. Simple Lipids

Esters of fatty acids with various alcohols. It contains two parts:

Fats: They are found in nature in large quantities. Fats are the best reserve of food material in the human body. Fats consist of a wide group of compounds that are generally soluble in organic solvents and largely insoluble in water. They are esters of fatty acids with glycerol. Fats may be either solid or liquid at room temperature, depending on their structure and composition. Although the words "oils", "fats", and "lipids" are all used to refer to fats, "oils" is usually used to refer to fats that are liquids at normal room temperature, while "fats" is usually used to refer to fats that are solids at normal room temperature. "Lipids" is used to refer to both liquid and solid fats, along with other related substances. The word "oil" is used for any substance that does not mix with water and has a greasy feel, such as petroleum (or crude oil) and heating oil, regardless of its chemical structure. Fats form a category of lipid, distinguished from other lipids by their chemical structure and physical properties. This category of molecules is important for many forms of life, serving both structural and metabolic functions. They are an essential part of the diet of most heterotrophs (including humans). The enzyme is responsible to break down the fats or lipids molecules in the body is called lipases; it is produced in the pancreas. Lipases in the intestine assist in the digestion and absorption of dietary fat. Adipocytes and germinating seeds contain lipases that break down stored triglycerols, releasing fatty acids for export to other tissues, where they are required as fuel. Examples of edible animal fats are lard (pig fat), fish oil, and butter or ghee. They are obtained from fats in the milk, meat and under the skin of the animal. Vegetable oils, sunflower, soya bean, sesame, coconut, olive, and Peanut are the examples of edible plant fats.

Chemical Structure of Fat

All fats consist of fatty acids (chains of carbon and hydrogen atoms, with a carboxylic acid group at one end) bonded to a backbone structure, often glycerol (a "backbone" of carbon, hydrogen, and oxygen). Chemically, this is a triester of glycerol, an ester being the molecule formed from the reaction of the carboxylic acid and an organic alcohol. The properties of any specific fat molecule depend on the particular fatty acids that constitute it. Different fatty acids are composed of different numbers of carbon and hydrogen atoms. The carbon atoms, each bonded to two neighbouring carbon atoms, form a zigzagging chain; the more carbon atoms there are in any fatty acid, the longer its chain will be. Fatty acids with long chains are more susceptible to intermolecular forces of attraction (in this case, Vander Waals forces), raising its melting point. Long chains also yield more energy per molecule when metabolized.

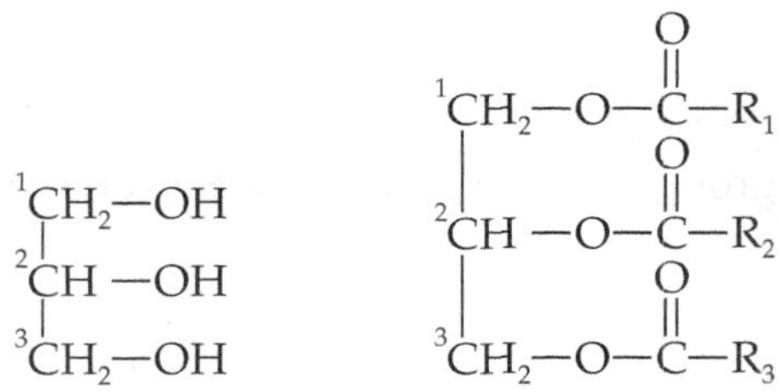

Glycerol Triacylglycerol

Fig. 2.34 Structure of Glycerol and Triacylglycerol

Adipose Tissue: Function as Energy Reservoir

In animals, adipose, or fatty tissue is the body's means of storing metabolic energy over extended periods of time. Depending on current physiological conditions, adipocytes (Fig. 2.35) (a) store fat derived from the diet and liver metabolism or degrade stored fat to supply fatty acids and glycerol to the circulation. These metabolic activities are regulated by several hormones (*i.e.*, insulin, glucagon and epinephrine). The location of the tissue determines its metabolic profile: "Visceral fat" is located within the abdominal wall (*i.e.*, beneath the wall of abdominal muscle) whereas "subcutaneous fat" is located beneath the skin (and includes fat that is located in the abdominal area beneath the skin but above the abdominal muscle wall). Visceral fat was recently discovered to be a significant producer of signalling chemicals (*i.e.* hormones), among which are several which are involved in inflammatory tissue responses. One of these is resist_in which has been linked to obesity, insulin resistance, and Type 2 diabetes.

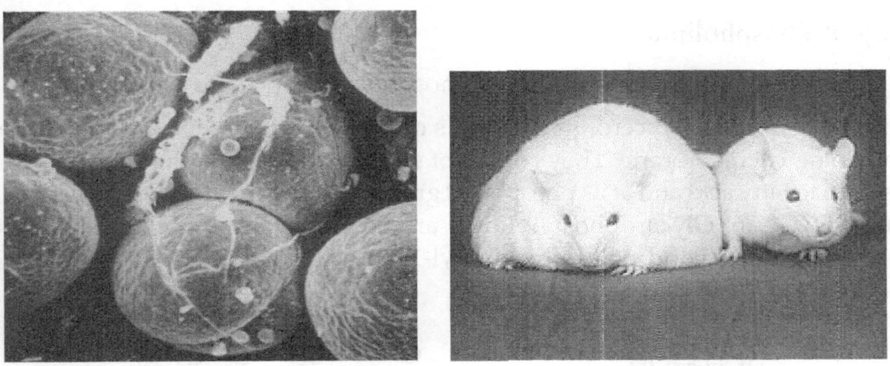

(a) (b)

Fig. 2.35 (*a*) Scanning electron micrograph of adipocytes. (*b*) The obese mouse on the left has large stores of adipose tissue. For comparison, a mouse with a normal amount of adipose tissue is shown on the right.

Waxes

Waxes are esters of long chain fatty acids combined with long chain monohydric alcohols. Although devoid of glycerol esters, they have only one ester linkage. Waxes are mainly used as a waterproofing material by plants and animals, *e.g.*, on leaves of plants and on the feathers of animals. Waxes are mainly three types.

(*a*) True waxes are esters of higher fatty acids with acetyl alcohol or other higher straight chain alcohols.

(*b*) Cholesterol esters are esters of fatty acid with cholesterol.

(*c*) Vitamin A and vitamin D esters are palmitic or stearic acid esters of vitamin A or vitamin D respectively.

2. Compound Lipids

Esters of fatty acids containing groups in addition to an alcohol and a fatty acid. It is grouped under following section:

Phospholipids

Glycolipids

Lipoproteins

Phospholipids

Phospholipids are lipids containing a phosphate group, they are found in cellular membranes. They are present in large amount in nerve tissue, brain, liver, kidney, pancreas and heart.

Biological Functions of Phospholipids

1. They increase the rate of fatty acid oxidation.
2. They act as carriers of inorganic ions across the membrane.
3. They help in blood clotting.
4. They act as a prosthetic group to certain enzyme.
5. They form the structures of membrane, matrix of cell wall, myelin sheath, microsomes and mitochondria.

Classification of Phospholipids

It is based on the type of alcohol present in the phospholipid. They are three types:

(*a*) ***Glycerophospholipds:*** Glycerophospholipds or phosphoglycerides. The major lipid component of biological membrane. They consist of **sn-glycerol-3-phosphate** esterified at it C_1 and C_2 positions to fatty acid and at its phosphoryl group to a group, X, to form the class of substances shown in Fig. 2.36. Glycerophospholipids are therefore amphiphilic molecules with non polar aliphatic "tails" and polar phosphoryl-X "heads".

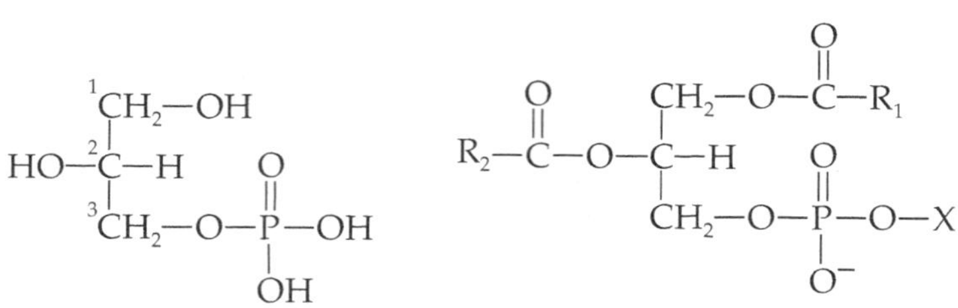

sn-Glycerol-3-phosphate
(a)

Glycerophospholipid
(b)

Fig. 2.36 (*a*) Structure of glycerol-3-phosphate (*b*) The general formula of the glycerophospholipids. R_1 and R_2 are long-chain hydrocarbon tails of fatty acids and X is derived from a polar alcohol.

Common classes of Glycerophospholipids (Table 2.14).

(*i*) Phosphatidyl ethanolamine (cephalin)
(*ii*) Phosphatidyl choline (lecithin)
(*iii*) Phosphatidyl serine
(*iv*) Plasmalogens
(*v*) Phosphatydic acid

Table 2.14 The common classes of Glycerophospholipids

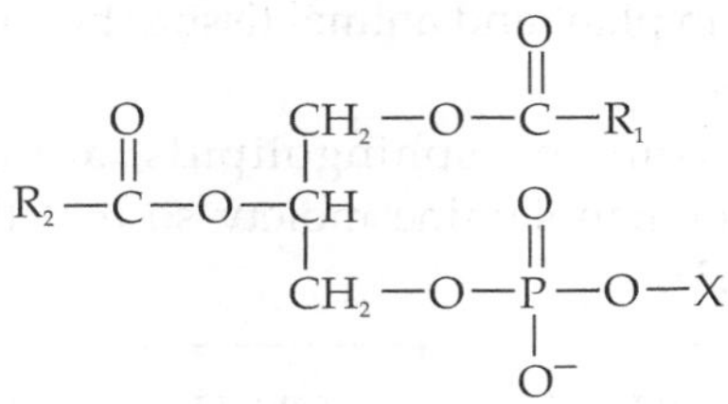

Name of X — OH	Formula of — X	Name of Phospholipid
Water	$-H$	Phosphatidic acid
Ethanolamine	$-CH_2CH_2NH_3^+$	Phosphatidylethanolamine
Choline	$-CH_2CH_2N(NH_3)_3^+$	Phosphatidylcholine (lecithin)
Serine	$-CH_2CH(NH_3^+)COO^-$	Phosphatidylserine
myo-Inositol	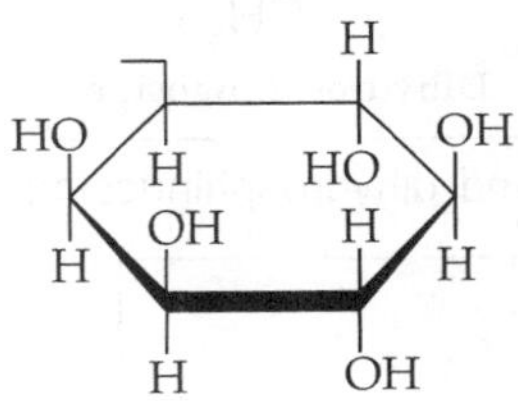	Phosphatidylinositol
Glycerol	$-CH_2CH(OH)\ CH_2OH$	Phosphatidylglycerol
Phosphatidylglycerol	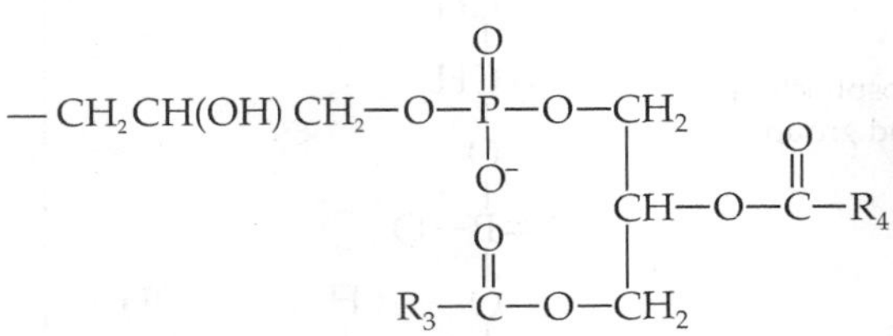	Diphosphatidylglycerol (cardiolipin)

(*b*) ***Phosphoinositides:*** In this, inositol is the alcohol. *e.g.*, Phosphatidyl inositol (lipositol).

(*c*) ***Phosphosphingosides:*** sphingolipids are major membrane components, derivatives of the C_{18} amino alcohols sphingosine, dihydrosphingosine (Fig. 2.37) and there C_{16}, C_{17}, C_{19} and C_{20} homologs. There N-acetyl fatty acid derivatives ceramides,

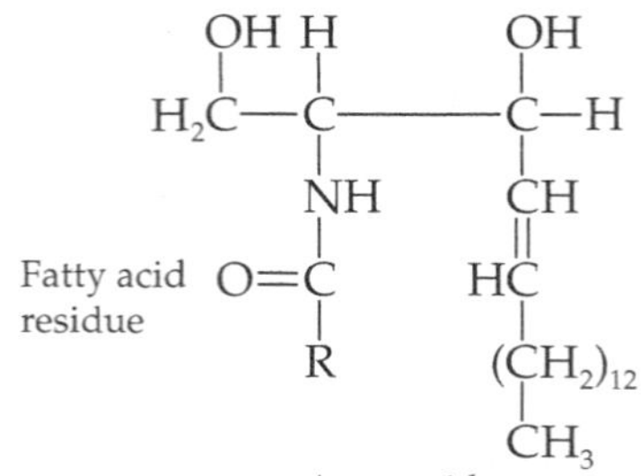

A ceramide

Occurs only in small amounts in plant and animal tissues but form the parent compounds of more abundant sphingolipids:

Sphingomyelins, the most common sphingolipids, are ceramides bearing either a phosphocholine or a phosphoethanolamine moiety so that they can also be classified as sphingolipids shown in Fig. 2.38.

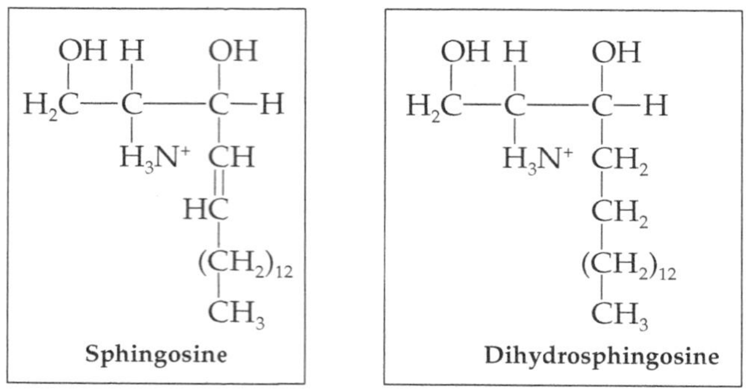

Fig. 2.37 Structure of sphingosine and dihydrosphingosine

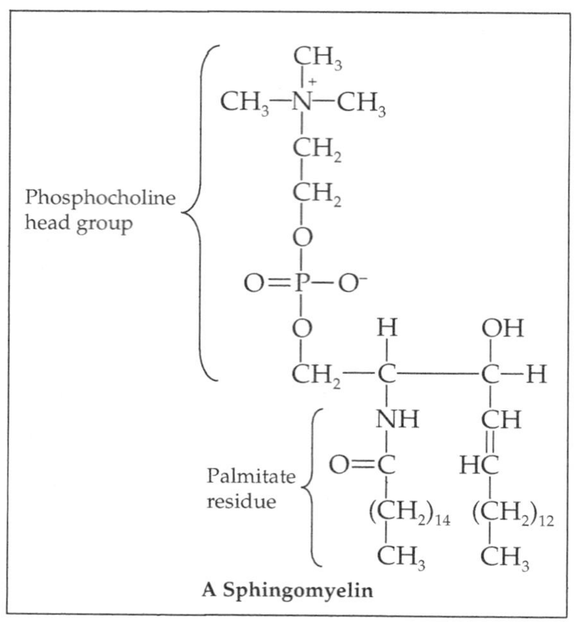

Fig. 2.38 Structure of sphingomyelin

Glycolipids

These contain an amino alcohol (sphingosine or isosphingosine) attached with an amide linkage to a fatty acid and glycosidically to a carbohydrate moiety. These are further classified in to two groups.

(*a*) ***Cerebrosides:*** It contains galactose, a high molecular weight fatty acid and sphingosine. Therefore, they may also be classified as sphingolipids. ***Galactocerebrosides,*** Fig. 2.39 which are most prevalent in the neuronal cell membrane of the brain, have a β-D-galactose head group. ***Glucocerebrosides,*** which instead have a β-D-glucose residue, occurs in the membrane of other tissues.

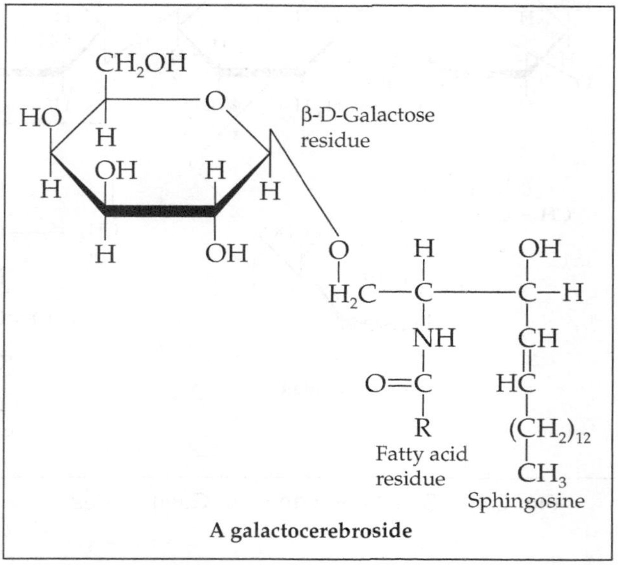

Fig. 2.39 Structure of Galactocerebrosides

(*b*) ***Gangliosides:*** They are ceramide oligosachharides that include among their sugar group atleast one sialic acid residue (N-acetylneuraminic acid). The structures of Gangliosides G_{M1}, G_{M2} and G_{M3}, three of the over sixty that are known, shown in Fig. 2.40. Gangliosides are primarily components of cell surface membrane and constitute a significant fraction (6%) of brain lipids. Gangliosides are also receptors fro bacterial protein toxins such as cholera toxin. Gangliosides have considerable physiological and medical significance. Disorders of Gangliosides break down are responsible for several hereditary sphingolipid storage diseases, such as Tay-Sachs disease, which are characterized by invariably fatal neurological deterioration.

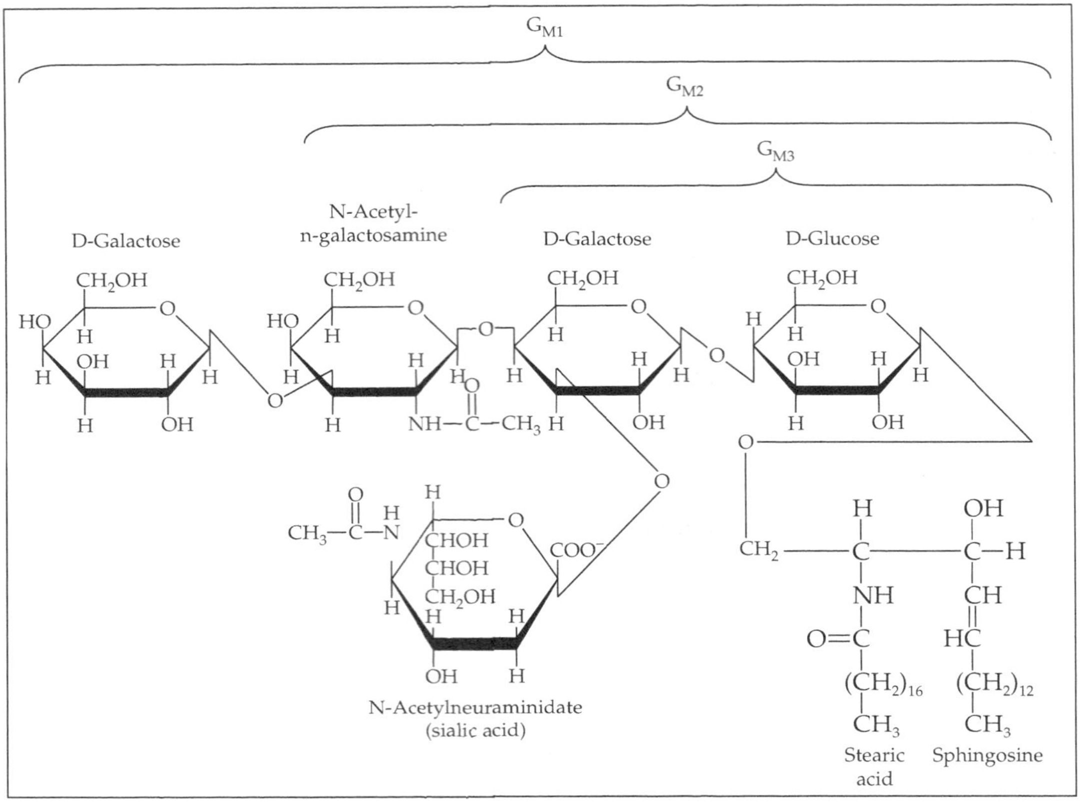

Fig. 2.40 Structural formula of Gangliosides

Lipoproteins

Lipoproteins are globular micelle like particles that consist of a non polar core of triacylglycerols and cholesteryl esters surrounded by an amphiphilic coating of protein, phospholipid and cholesterol. Lipoproteins are combinations of lipids and proteins, they are found in membranes. When lipids are transported through the blood and lymphatic systems, they are in the form of lipoproteins. Five major groups of Lipoproteins have been identified which are important physiologically and in clinical diagnosis in some metabolic disorder of fat metabolism. These are:

(*a*) Chylomicrons
(*b*) Very low density lipoprotein (VLDL or pre-β-lipoprotein)
(*c*) Intermediate density lipoprotein (IDL)
(*d*) Low density lipoprotein (LDL or b-lipoprotein)
(*e*) High density lipoprotein (HDL or a-lipoprotein)

(*a*) **Chylomicrons,** which transport exogenous triacylglycerols and cholesterol from the intestine to the tissues.

(*b-d*). **VLDL, IDL, LDL,** a group of related particles that transport endogenous triacylglycerols and cholesterol form liver to the tissues (the liver synthesises triacylglycerols from excess carbohydrates).

(*e*) **HDL,** which transport endogenous cholesterol from the tissues to the liver.

Lipoprotein particles under the continuous metabolic processing so that they have somewhat variable properties and compositions Table 2.15. Each contains just enough protein, phospholipid and cholesterol to form an ~20 Å thick monolayer of these substances on the particle surface. Lipoprotein densities increase with decreasing particle diameter because the density of their outer coating is greater than that of their inner core.

Table 2.15 Description of the major classes of lipoproteins in human plasma

	Chylomicrons	VLDL	IDL	LDL	HDL
Density(g.cm^{-3})	<0.95	<1.006	1.006–1.019	1.019–1.0	1.063–1.210
Particle diameter (Å)	750–12,000	300–800	250–350	180–250	50–120
Particle mass (kD)	400,000	10–80,000	5–10,000	2300	175–360
% protein[a]	1.5–2.5	5–10	15–20	20–25	40–55
% phospholipids[a]	7–9	15–20	22	15–20	20–35
% free cholesterol[a]	1–3	5–10	8	7–10	3–4
% triacylglycerols[b]	84–89	50–65	22	7–10	3–5
% cholesterylesters[b]	3–5	10–15	30	35–40	12
Major apolipoproteins*	A-I, A-II, B-48, C-I, C-II, C-III, E	B-100,C-I, C-II, C-III, E	B-100, C-III,E	B-100	A-I, A-II, C-I, C-II, C-III, D ,E

[(a)]surface components
[(b)]core lipids
The protein components of lipoproteins are known as apolipoproteins or just apoproteins.

Importance of Lipoprotein

- To maintain structural integrity of cell surface and sub cellular particles like mitochondria and microsomes.
- To transport and deliver the lipids to tissues.
- The β-lipoprotein fraction increases in severe diabetes mellitus, atherosclerosis etc.

3. Derived Lipids

These include fatty acids, glycerol, steroids, cholesterol, other alcohols, fatty aldehydes and ketone bodies, hydrocarbons, lipid soluble vitamins and hormones.

A. Fatty Acids

Fatty acids are carboxylic acids with long chain hydrocarbon side groups shown in Fig. 2.41. They are rarely free in nature but, rather, occur in esterified form as the major components of the various lipids.

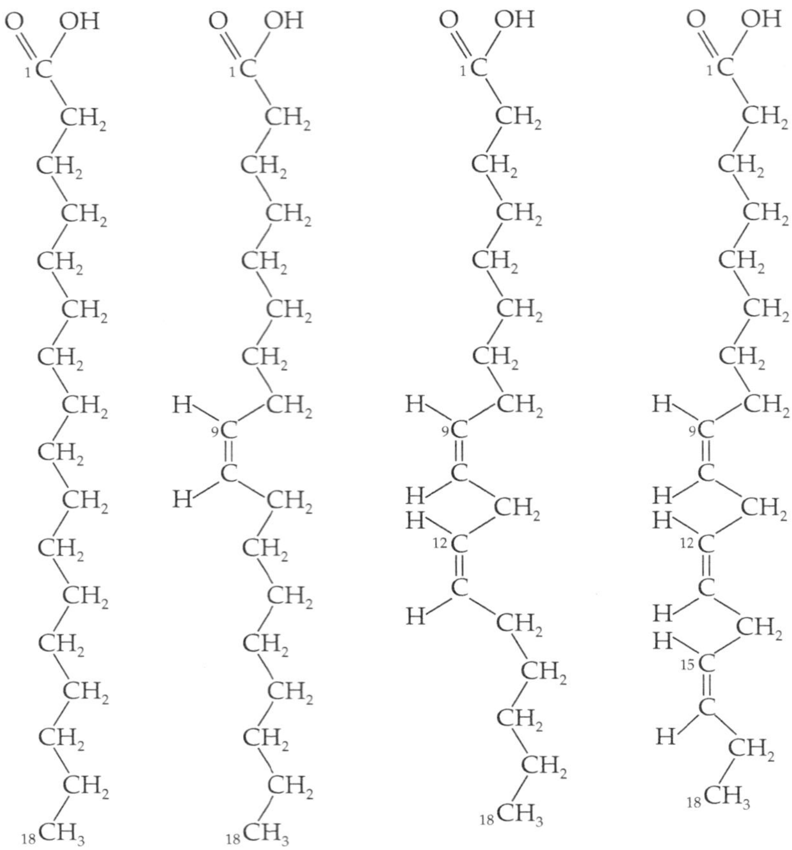

Fig. 2.41 Structural formula of some C_{18} fatty acids

Properties of Fatty Acids

1. These are obtained by the hydrolysis of fats.
2. Fatty acids occurring in natural fats usually contain an even number of carbon atoms because they are synthesized from two carbon units and are straight chain derivatives.
3. The straight chain may be saturated (containing no double bonds) or unsaturated (containing one or more double bonds) as shown in Table 2.16.
4. Carbon atoms of fatty acids are numbered from the carboxyl carbon (carbon number-1) the carbon atom adjacent to the carboxyl carbon (carbon number-2) is also known as the α-carbon.

Saturated Fatty Acids

Saturated fatty acids have long hydrocarbon chains terminating in a carboxyl group. The general formula of a saturated fatty acid is $CH_3-(CH_2)n-COOH$. In naturally occurring fatty acids, the n referred to in the general formula is normally an even number and is somewhere between 2 and 22.

Unsaturated Fatty Acids

In unsaturated fatty acids, at least two and no more than six carbon atoms in the hydrocarbon chain are linked together by double bonds.

Table 2.16 The common biological fatty acids

Symbol[a]	Common Name	Systematic Name	Structure
		Saturated fatty acids	
12 : 0	Laurie acid	Dodecanoic acid	$CH_3(CH_2)_{10}COOH$
14 : 0	Myristic acid	Tetradecanoic acid	$CH_3(CH_2)_{12}COOH$
16 : 0	Palmitic acid	Hexadecanoic acid	$CH_3(CH_2)_{14}COOH$
18 : 0	Stearic acid	Octadecanoic acid	$CH_3(CH_2)_{16}COOH$
20 : 0	Arachidic acid	Eicosanoic acid	$CH_3(CH_2)_{,8}COOH$
22 : 0	Behenic acid	Docosanoic acid	$CH_3(CH_2)_{20}COOH$
24 : 0	Lignoceric acid	Tetracosanoic acid	$CH_3(CH_2)_{22}COOH$
		Unsaturated fatty acids (all double bonds are cis)	
16 : 1	Palmitoleic acid	9-Hexadecenoic acid	$CH_3(CH_2)_5CH{=}CH(CH_2)_7COOH$
18 : 1	Oleic acid	9-Octadecenoic acid	$CH_3(CH_2)_7CH{=}CH(CH_2)_7COOH$
18 : 2	Linoleic acid	9,12-Octadecadienoic acid	$CH_3(CH_2)_4(CH{=}CHCH_2)_2(CH_2)_6COOH$
18 : 3	α-Linolenic acid	9,12,15-Octadecatrienoic acid	$CH_3CH_2(CH{=}CHCH_2)_3(CH_2)_6COOH$
18 : 3	γ-Linolenic acid	6,9,12-Octadecatrienoic acid	$CH_3(CH_2)_4(CH{=}CHCH_2),(CH_2)_3COOH$
20 : 4	Arachidonic acid	5,8,11,14-Eicosatetraenoic acid	$CH_3(CH_2)_4(CH{=}CHCH_2)_4(CH_2)_2COOH$
20 : 5	EPA	5,8,11,14,17-Eicosapentanoic acid	$CH_3CH_2(CH{=}CHCH_2)_5(CH_2)_2COOH$
24 : 1	Nervonic acid	15-Tetracosenoic acid	$CH_3(CH_2)_7CH{=}CH(CH_2)_{13}COOH$

[a] Number of carbon atoms : Number of double bonds.

Classification of Some Important Unsaturated Fatty Acids

Few important unsaturated fatty acids like Linolenic acid and arachidonic acid further united and give rise to eicosanoid. These eicosanoid further divided in to two parts *i.e.,* prostanoid and leukotriene. Prostanoid consists of prostaglandin and thromboxane Fig. 2.42.

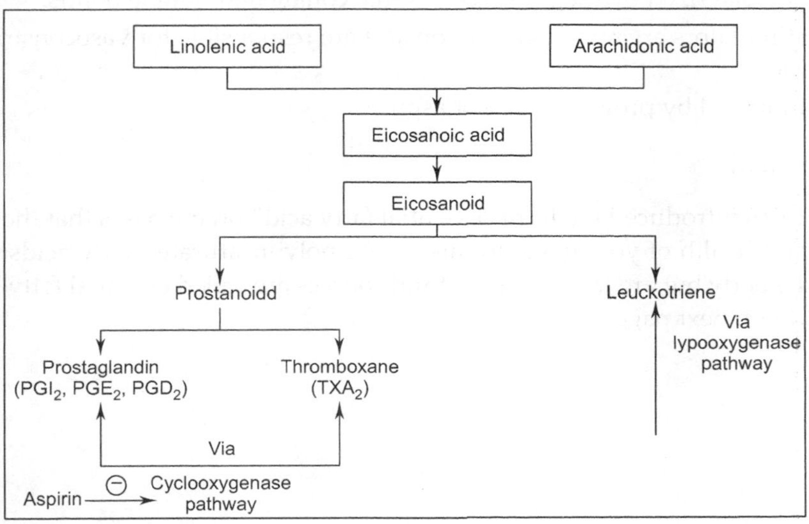

Fig. 2.42

Prostaglandin (PG)

1. They virtually exist in every mammalian tissue and act as a local hormones.
2. They have important physiologic and pharmacologic activities.
3. They are synthesized in vivo by cyclization of the centre of the carbon chain of 20 carbon polyunsaturated fatty acid to form a cyclopentane ring.
4. Three different eicosanoic fatty acids give rise to 3 groups of eicosanoids characterized by the number of double bonds in the side chain for example PG_1, PG_2 and PG_3.

Prostacyclins (PGI)

1. They are formed in vascular endothelium and continually form in heart. They are also formed in kidney.
2. They are formed from cyclic endoperoxide PGH_2 by the action of microsomal prostacyclin synthetase.
3. They inhibit platelet aggregation and gastric secretion from the pyloric mucosa.
4. They decrease blood pressure and protect coronary arteries.
5. They increased renal blood flow and stimulate rennin production.

Thromboxane

1. They contract smooth muscles on blood vessels, uterus.
2. They are discovered in platelets, and have the cyclopentane ring interrupted with an oxygen atom (oxane ring).
3. The substituent groups attached to the rings being varied give rise to different type in each series of thromboxane levelled A, B etc.
4. They produce vasoconstriction and increase blood pressure.
5. Imidazoles inhibit their synthesis.

Leukotrienes

1. They are characterized by the presence of three conjugated double bonds.
2. They are stimulators are mucous secretion and are responsible for vasoconstriction of bronchial muscles.
3. They are inhibited by prolonged use of aspirin.

Essential Fatty Acids

Burr and Burr (1930) introduced the term "essential fatty acid" on the bases that they are essential for the growth and health of young albino rats. These polyunsaturated fatty acids which are not synthesized in the body but are taken from natural sources are called essential fatty acid. They are listed in Table 2.17 on next page.

Table 2.17

Essential fatty acids	Chemical name	General formula	No. of double bonds	Structure	Sources
Linoleic	9,12-Octa-decadienoic acid	$C_nH_{2n-3}COOH$	$2(\Delta^9, \Delta^{12})$	$CH_3-(CH_2)_4$ $(CH=CH-CH_2)_2$ $(CH_2)_6COOH$	Corn, Peanut, Cotton seed, Soyabean oil.
Linolenic	6,9,12-Octa-decatrienoic acid	$C_nH_{2n-5}COOH$	$3(\Delta^9, \Delta^{12}, \Delta^{15})$	$CH_3-(CH_2)_4$ $(CH=CH-CH_2)_3$ $(CH_2)_3COOH$	Found frequently with linoleic acid but particularly in linseed oil.
Arachidonic	5,8,11,14-eico-satetraenoic acid	$C_nH_{2n-7}COOH$	$4(\Delta^5, \Delta^8, \Delta^{11}, \Delta^{14})$	$CH_3-(CH_2)_4$ $(CH=CH-CH_2)_4$ $(CH_2)_2COOH$	Found in small quantities with linoleic and linolenic acids but particularly in peanut oil.

B. Refined and Hydrogenated Oil

Refind oil is prepared in the following manner:

1. Free fatty acids are removed by alkali treatment.
2. Colouring mater is removed by activated carbon.
3. Odour is removed by super heated steam.

Hydrogenated Oil

The refined oils are hydrogenated under optimum temperature and pressure with hydrogen in the presence of nickel catalyst. Unsaturated fatty acids are converted into saturated fatty acids. Vanaspati is hydrogenated refined ground nut oil.

C. Steroids

The steroids are often found in association with fat. They have a similar cyclic nucleus resembling phenanthrene (rings A, B, C) to which a cyclopentane ring (D) is attached. The parent substance is better designated as cyclopentane-per-hydrophenanthrene. Methyl side chain occurs typically at positions 10 and 13. A side chain at position 17 is usual (as in cholesterol). If the compound has one or more hydroxyl groups and no carbonyl or carboxyl groups, it is a sterol. Steroids may be divided in the following manner.

(*i*) Sterols: cholesterol, Ergosterol, coprosterol
(*ii*) Bile acids: Glycocholic acid and taurocholic acid
(*iii*) Sex hormones: Testosterone, estradiol
(*iv*) Vitamin D: Vit.D_2 and D_3
(*v*) Adrenocortical hormones: Corticosterone
(*vi*) Cardiac glycosides: Stropanthin
(*vii*) Saponins: Digitonin

Cholesterol

Cholesterol is a major component of animal plasma membranes and occurs in lesser amounts in the membranes of their sub cellular organelles. Its polar —OH group gives it a weak amphiphilic character, where as its fused ring system provides it with greater rigidity than other membrane lipids. Cholesterol is the metabolic precursor of steroid hormones. The structure of parent compound of steroid and Cholesterol are given below Fig. 2.43 (a and b):

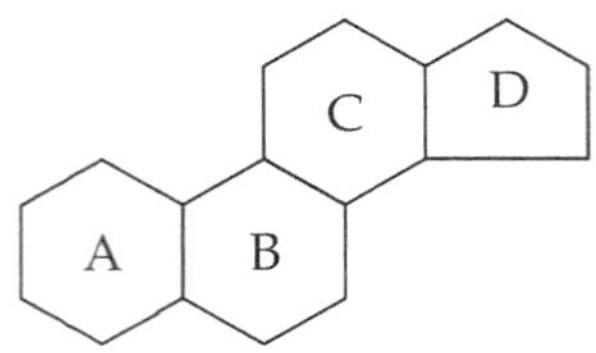

Cyclopentanoperhydrophenanthrene

Fig. 2.43 (*a*) The parent compound of steroid consists of four fused saturated rings.

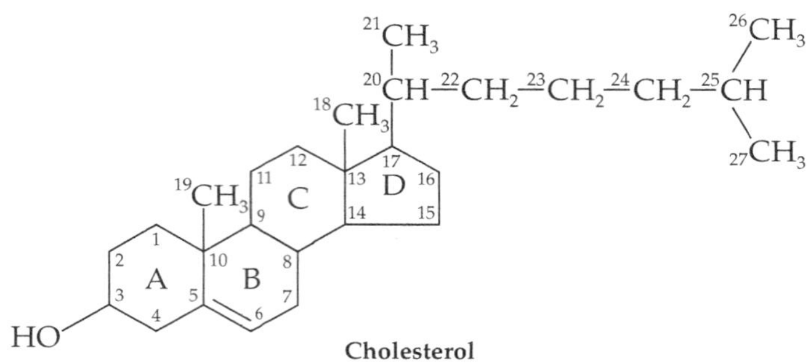

Cholesterol

Fig. 2.43 (*b*) Structural formula with the standard numbering system

Ergosterol

Yeast and fungi have other membrane sterols such as **ergosterol,** which have C7 to C8 double bond (Fig. 2.44).

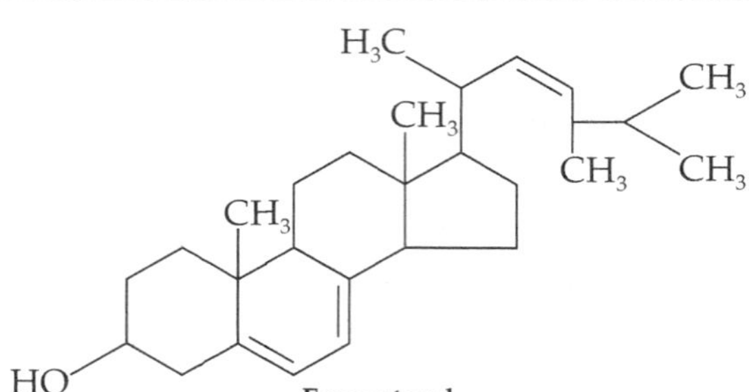

Erogosterol

Fig. 2.44 Structural formula of ergosterol

Biological Functions of Lipids

Having discussed at some length the chemistry of lipids, their mode of occurrence in tissues and their state of dynamic equilibrium, it would be satisfying if the story could be nearly rounded off by showing the connexion between these features and the diverse functions of various lipid classes in living organisms.

1. ***Membranes:*** Eukaryotic cells are compartmentalized into membrane-bound organelles which carry out different biological functions. The glycerophospholipids are the main structural component of biological membranes, such as the cellular plasma membrane and the intracellular membranes of organelles; in animal cells the plasma membrane physically separates the intracellular components from the extracellular environment. The glycerophospholipids are amphipathic molecules (containing both hydrophobic and hydrophilic regions) that contain a glycerol core linked to two fatty acid-derived "tails" by ester linkages and to one "head" group by a phosphate ester linkage. While glycerophospholipids are the major component of biological membranes, other non-glyceride lipid components such as sphingomyelin and sterols (mainly cholesterol in animal cell membranes) are also found in biological membranes. In plants and algae, the galactosyldiacylglycerols, and sulfoquinovosyldiacylglycerol, which lack a phosphate group, are important components of membranes of chloroplasts and related organelles and are the most abundant lipids in photosynthetic tissues, including those of higher plants, algae and certain bacteria.

   A biological membrane is a form of lipid bilayer. The formation of lipid bilayers is an energetically-preferred process when the glycerophospholipids described above are in an aqueous environment. In an aqueous system, the polar heads of lipids align towards the polar, aqueous environment, while the hydrophobic tails minimize their contact with water and tend to cluster together, forming a vesicle; depending on the concentration of the lipid, this biophysical interaction may result in the formation of micelles, liposomes, or lipid bilayers. Other aggregations are also observed and form part of the polymorphism of amphiphile (lipid) behavior. Phase behavior is an area of study within biophysics and is the subject of current academic research. Micelles and bilayers form in the polar medium by a process known as the hydrophobic effect. When dissolving a lipophilic or amphiphilic substance in a polar environment, the polar molecules (*i.e.*, water in an aqueous solution) become more ordered around the dissolved lipophilic substance, since the polar molecules cannot form hydrogen bonds to the lipophilic areas of the amphiphile. So in an aqueous environment, the water molecules form an ordered "clathrate" cage around the dissolved lipophilic molecule.

2. ***Energy storage:*** Triacylglycerols, stored in adipose tissue, are a major form of energy storage in animals. The adipocyte, or fat cell, is designed for continuous synthesis and breakdown of triacylglycerols, with breakdown controlled mainly by the activation of hormone-sensitive enzyme lipase. The complete oxidation of fatty acids provides high caloric content, about 9 kcal/g, compared with 4 kcal/g for the breakdown of carbohydrates and proteins. Migratory birds that must fly long distances without eating use stored energy of triacylglycerols to fuel their flights.

3. ***Signalling:*** In recent years, evidence has emerged showing that lipid signalling is a vital part of the cell signalling. Lipid signalling may occur via activation of G protein-coupled or nuclear receptors, and members of several different lipid categories have been identified as signalling molecules and cellular messengers. These include sphingosine-1-phosphate, a sphingolipid derived from ceramide that is a potent messenger molecule involved in regulating calcium mobilization, cell growth, and apoptosis; diacylglycerol (DAG) and the phosphatidylinositol

phosphates (PIPs), involved in calcium-mediated activation of protein kinase C; the prostaglandins, which are one type of fatty-acid derived eicosanoid involved in inflammation and immunity; the steroid hormones such as estrogen, testosterone and cortisol, which modulate a host of functions such as reproduction, metabolism and blood pressure; and the oxysterols such as 25-hydroxy-cholesterol that are liver X receptor agonists.

4. ***Other functions:*** The "fat-soluble" vitamins (A, D, E and K), which are isoprene-based lipids; are essential nutrients stored in the liver and fatty tissues, with a diverse range of functions. Acyl-carnitines are involved in the transport and metabolism of fatty acids in and out of mitochondria, where they undergo beta oxidation. Polyprenols and their phosphorylated derivatives also play important transport roles, in this case the transport of oligosaccharides across membranes. Polyprenol phosphate sugars and polyprenol diphosphate sugars function in extra-cytoplasmic glycosylation reactions, in extracellular polysaccharide biosynthesis (for instance, peptidoglycan polymerization in bacteria), and in eukaryotic protein N-glycosylation. Cardiolipins are a subclass of glycerophospholipids containing four acyl chains and three glycerol groups that are particularly abundant in the inner mitochondrial membrane. They are believed to activate enzymes involved with oxidative phosphorylation.

2.2.4 Nucleic Acid

Nucleic acids are the polynucleotides; there are two types of Nucleic acids Deoxyribosnucleic acids and Ribonucleic acids. DNA is the chemical bases of heredity and organized in to genes that control every function, and composed of repeating units of Deoxyribonucleotides. The structure of every protein, and ultimately of every biomolecule and cellular component, is a product of information programmed in to the nucleotide sequence of a cell's nucleic acids. The ability to store and transmit genetic information from one generation to the next is a fundamental condition for life.

Components of Nucleic Acids

Nucleic acids are the polymers of nucleotides. In other words, nucleic acids are built up by the monomeric units of nucleotides.

Nucleotides

Nucleotides have three characteristics components:

1. A nitrogenous base
2. A pentose
3. A phosphate

The molecule without the phosphate group is called a nucleoside.

Biological Functions of Nucleotides

1. The nucleotides are important intracellular molecules of low molecular weight.
2. They are the building blocks of the nucleic acid.
3. They play an important role in carbohydrate, fat and protein metabolism.
4. Nucleotides are the structural components of several coenzymes of vitamin B-complex, *e.g.*, FAD, NAD^+, $NADP^+$, Coenzyme A.
5. The best role of purine and pyrimidine nucleotides is to serve as the monomeric precursors of RNA and DNA.

6. They serve as carriers of high energy intermediates in the biosynthesis of carbohydrate, lipid and protein; *e.g.*, UDP-glucose, CDP-acylglycerol, and UDP- galactose.
7. The purine nucleotides also act as the high energy source ATP.
8. Cyclic-AMP and cyclic-GMP are the second messenger in hormonal function.
9. Nucleotides control several metabolic reactions by their action as allosteric regulators.

Structure of Nucleotides

As already mentioned above that nucleotides consist of three components. The nitrogenous base consists of purines and pyrimidines. The structures of purines and pyrimidines bases are given following Fig. 2.45. There are two major purines found in living organisms:

(*a*) Adenine (6-aminopurine)
(*b*) Guanine (2-amino-6-oxypurine)

In plants, a series of purines bases containing methyl substituents occurs. Many have pharmacological properties. These include caffeine which is present in coffee, theophylline, which is present in tea, theobromine, which is present in coca. The structures are given below (Fig. 2.46).

There are three major pyrimidines found in living organisms:

(*a*) Cytosine (2-oxy-4-aminopyrimidine)
(*b*) Thymine (2,4-dioxy-5-methylpyrimidine)
(*c*) Uracil (2,4-dioxypyrimidine)

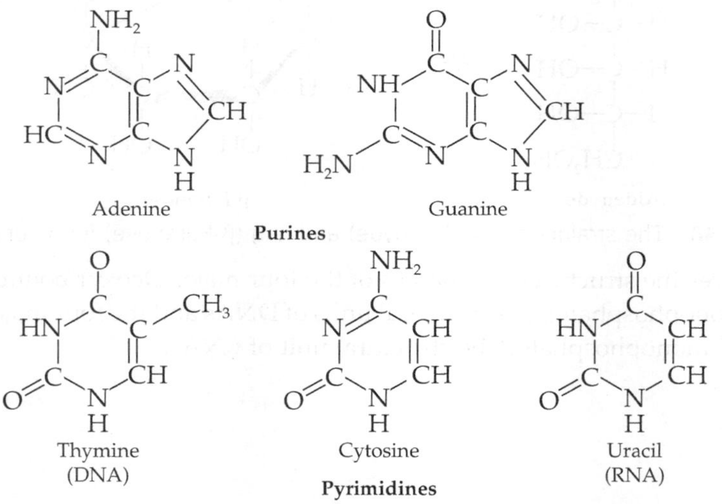

Fig. 2.45 The structures of purines and pyrimidines bases

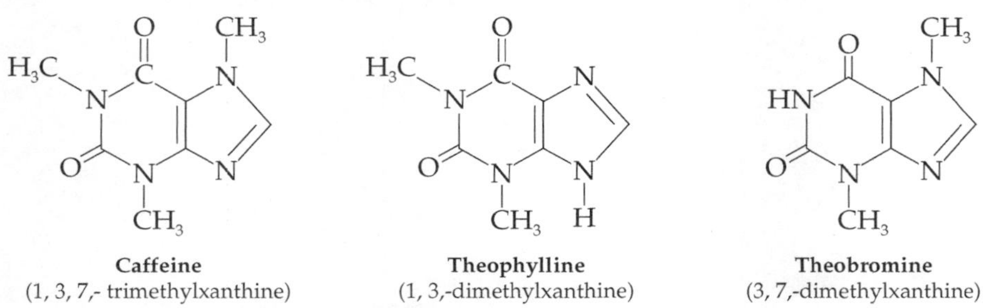

Fig. 2.46 The structures of plant purines

Tautomeric Forms of Purines and Pyrimidines

The existence of a molecule in a keto (lactam) and enol (lactin) form is known as tautomerism. Free purines and pyrimidines may exist two or more Tautomeric forms depending upon the pH. For example uracil occurs in lactam, lactin and double lactin form (Fig. 2.47):

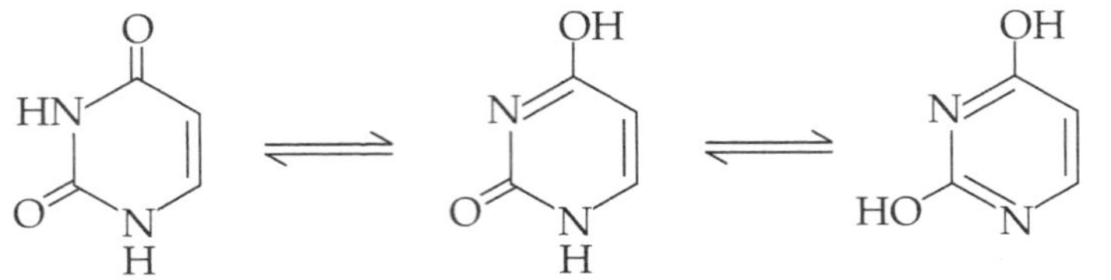

Fig. 2.47 Tautomeric forms of uracil

Sugars of Nucleic Acids

Nucleic acids have two types of pentoses. The deoxyribonucleotide units of DNA contain 2′-deoxy-D-ribose, and the Ribonucleotide units of RNA contain D-ribose. Both types of pentoses are in there b-furanose (closed 5 membered ring) form as shown in Fig. 2.48.

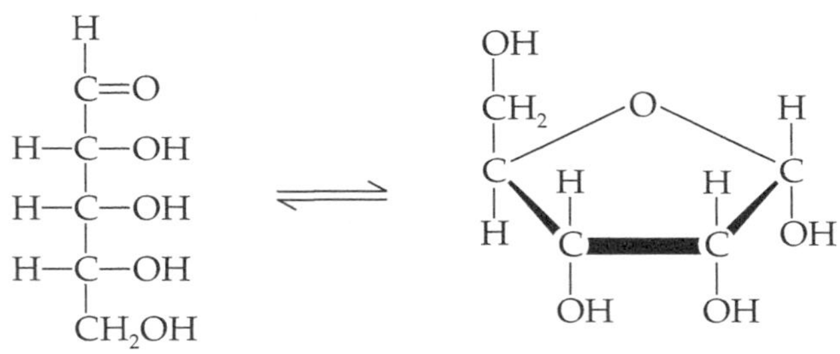

Fig. 2.48 The straight chain (Aldehyde) and ring (β-Furanose) forms of ribose

Figures 2.49, gives the structures and names of the four major Deoxyribonucleotides (Deoxyribonucleotide 5′-monophosphate), the structural units of DNAs and the four major Ribonucleotides (Ribonucleoside 5′-monophosphate), the structural unit of RNAs.

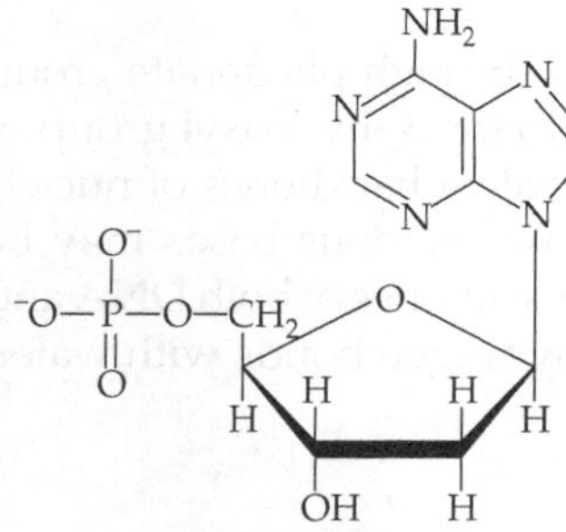

Nucleotide: Deoxyadenylate, deoxyadenosine 5′-monophosphate
Symbols: A, dA, dAMP
(Nucleoside: Deoxyadenosine)

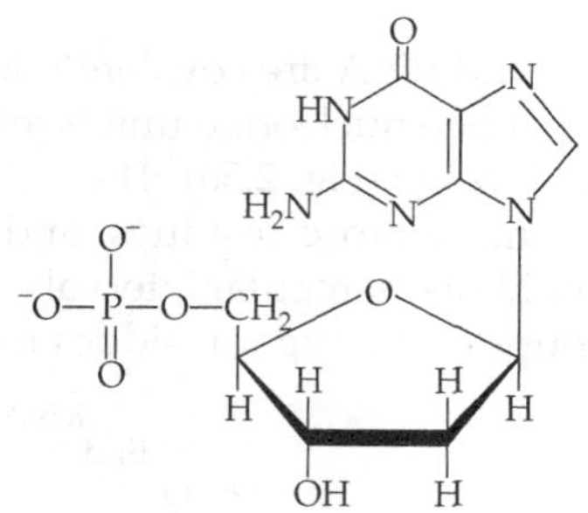

Deoxyguanylate, deoxyguanosine 5′-monophosphate
G, dG, dGMP
(Deoxguanosine)

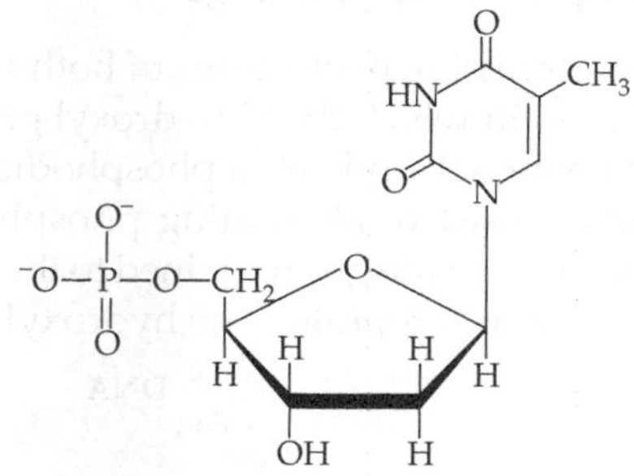

Deoxythymidylate, deoxythymidine 5′-monophosphate
T, dT, dTMP
(Deoxythymidine)

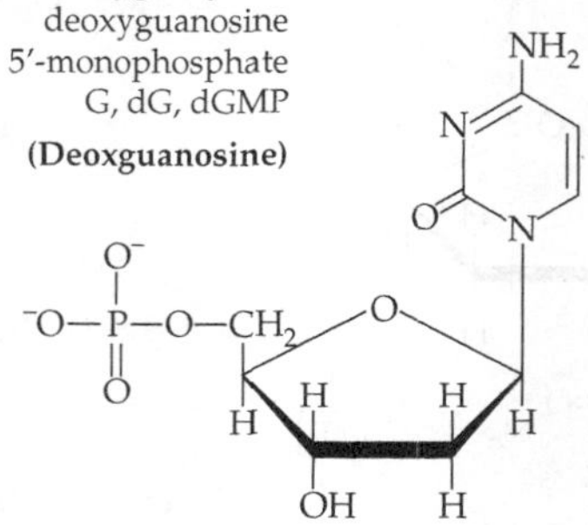

Deoxycylidylate, deoxycytidine 5′-monophosphate
C, dC, dCMP
(Deoxycytidine)

(a)

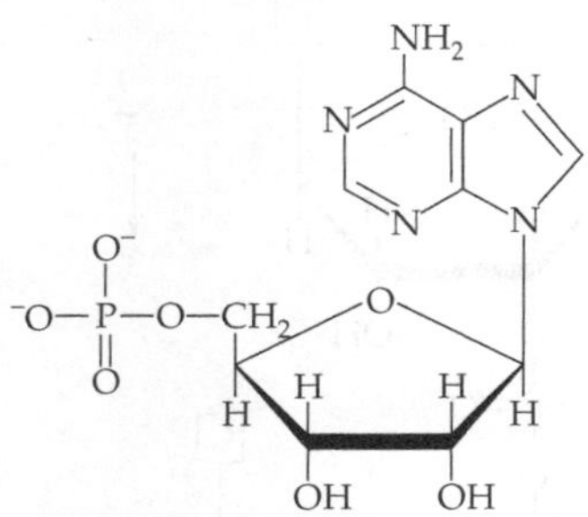

Nucleotide: Adenylate, adenosine 5′-monophosphate
Symbols: A, AMP
Nucleoside : (Adenosine)

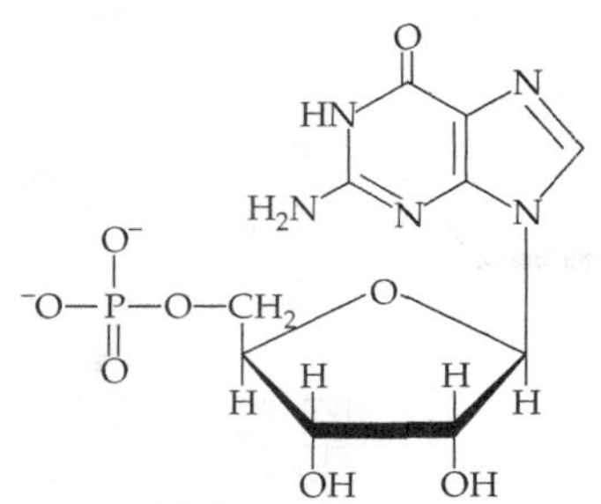

Guanylate, guanosine 5′-monophosphate
G, GMP
(Guanosine)

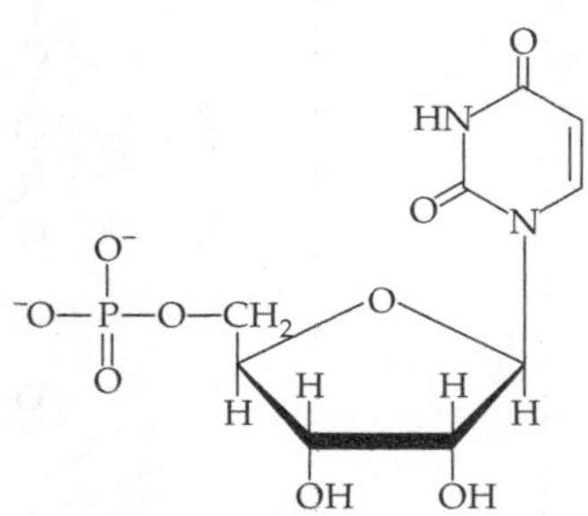

Uridylate, uridine 5′-monophosphate
U, UMP
(Uridine)

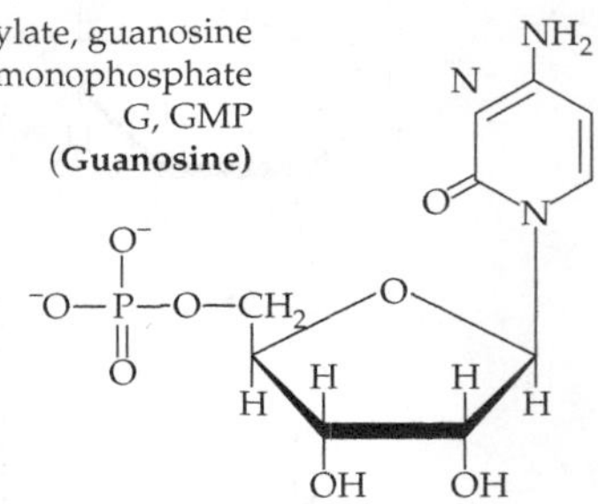

Cytidylate, cytidine, 5′-monophosphate
C, CMP
(Cytidine)

(b)

Fig. 2.49 *(a)* The deoxyribonucleotide units of DNA, *(b)* The ribonucleotide units of RNA

Phosphodiester Linkage

The successive nucleotides of both DNA and RNA are covalently linked through phosphate group bridges, in which the 5′-hydroxyl group of one nucleotide unit is joined to the 3′-hydroxyl group of the next nucleotide by a phosphodiester linkage (Fig. 2.50). Thus, the covalent backbones of nucleic acids consist of alternating phosphate and pentose residues, and the nitrogenous bases may be regarded as side groups joined to the e backbone at regular intervals. The backbones of both DNA and RNA are hydrophilic. The hydroxyl groups of the sugar residues form hydrogen bonds with water.

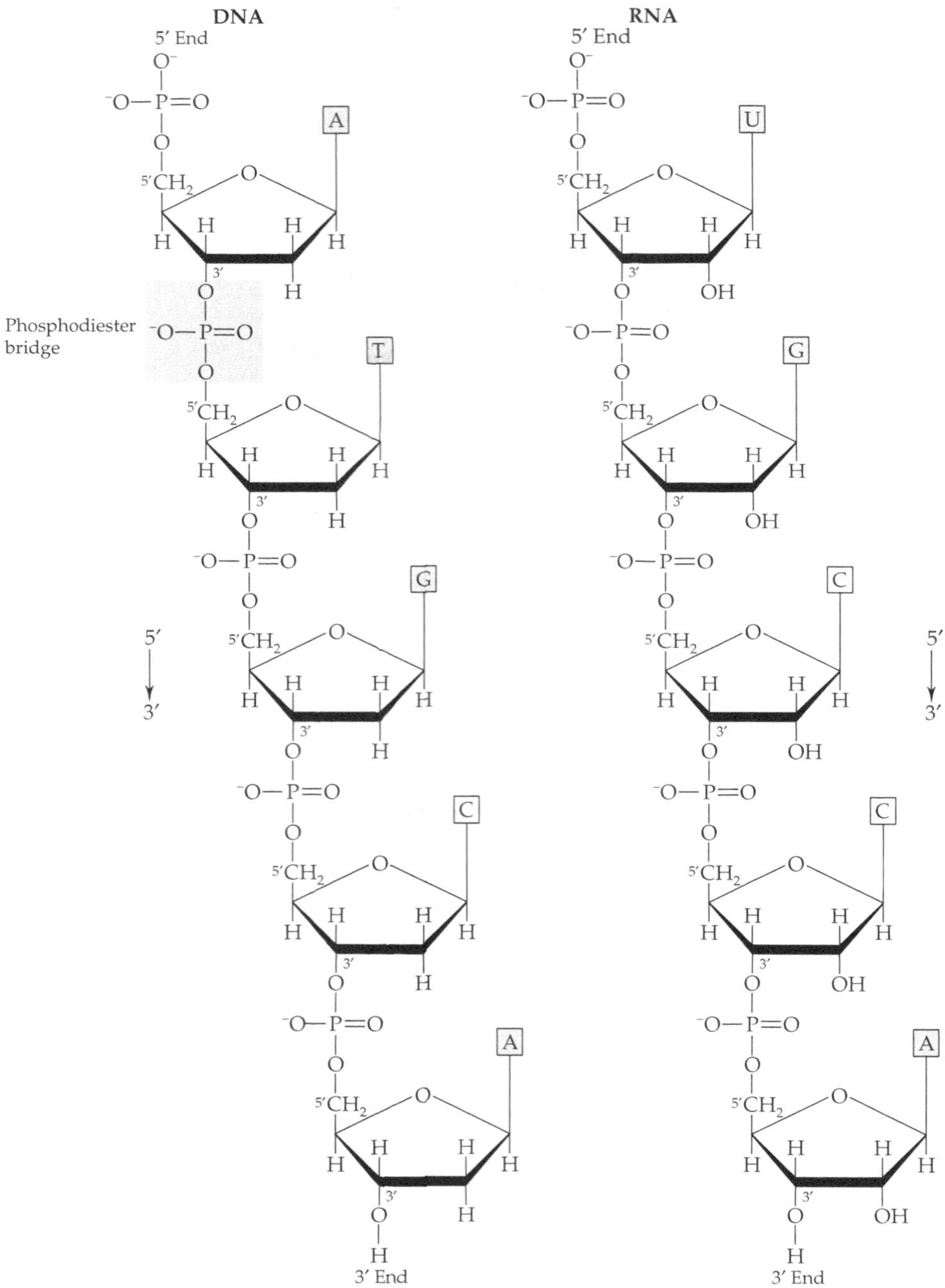

Fig. 2.50 The covalent backbone structure of DNA and RNA

Some Nucleotides are Work as Regulatory Molecules

One of the most common is adenosine 3′, 5′ cyclic monophosphate (cAMP), formed from ATP in a reaction catalysed by adenylate cyclise, an enzyme associated with the inner phase of the plasma membrane. cAMP serves regulatory functions in virtually every cell outside the plant kingdom. Guanosine 3′, 5′-cyclic monophosphate (cGMP) occurs in many cells and also has a regulatory function. Another regulatory nucleotide, ppGpp, is produced in bacteria in response to a slowdown in protein synthesis during amino acid starvation. This nucleotide inhibits the synthesis of the rRNA, and tRNA molecules needed for protein synthesis, preventing the unnecessary production of nucleic acids. Structures are given in following Fig. 2.51.

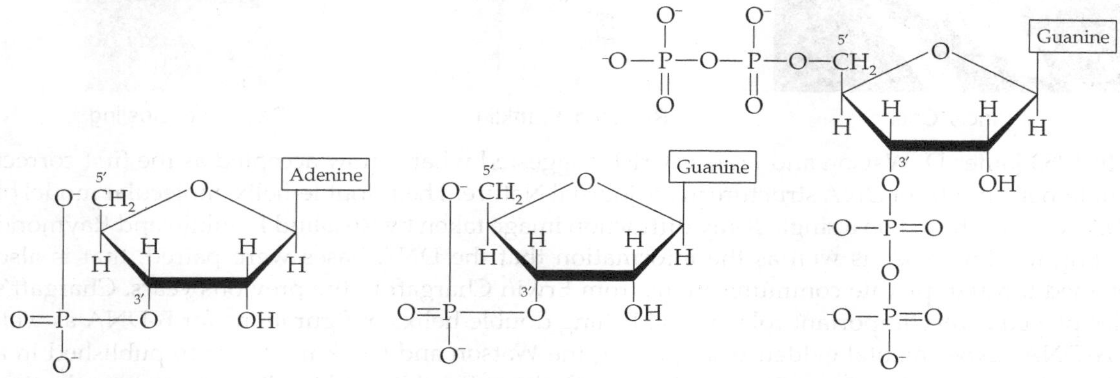

Adenosine 3′, 5′-cyclic monophosphate (cyclic AMP; cAMP) | Guanosine 3′, 5′-cyclic monophosphate (cyclic GMP; cGMP) | Guanosine 3′-diphosphate, 5′ -diphosphate (guanosine tetraphosphate) (ppGpp)

Fig. 2.51

Structure of Nucleic Acid

The structure of DNA was discovered by Watson and Crick in 1953. That was a momentous event in science, which gave rise to entirely new disciplines and influenced the course of many others. DNA is definitely the exclusive chromosomal component bearing the genetic information of living cell. The significance of this discovery appropriately recognise in 1962 when Watson, Crick and Wilkins receive the novel prize. Today, the fact that DNA is the genetic material is so fundamental to biological thought that it is difficult to realize what an enormous intellectual gap this discovery field.

Investigation History of DNA

DNA was first isolated by the Swiss physician Friedrich Miescher who, in 1869, discovered a microscopic substance in the pus of discarded surgical bandages. As it resided in the nuclei of cells, he called it "nuclein". In 1919, Phoebus Levene identified the base, sugar and phosphate nucleotide unit. Levene suggested that DNA consisted of a string of nucleotide units linked together through the phosphate groups. However, Levene thought the chain was short and the bases repeated in a fixed order. In 1937 William Astbury produced the first X-ray diffraction patterns that showed that DNA had a regular structure. In 1928, Frederick Griffith discovered that traits of the "smooth" form of the Pneumococcus could be transferred to the "rough" form of the same bacteria by mixing killed "smooth" bacteria with the live "rough" form. This system provided the first clear suggestion that DNA carried genetic information; the Avery-MacLeod-McCarty experiment—when Oswald Avery, along with co-workers Colin MacLeod and Maclyn McCarty, identified DNA as the transforming principle in 1943. DNA's role in heredity was confirmed in 1952, when Alfred Hershey and Martha Chase in the Hershey-Chase experiment showed that DNA is the genetic material of the T2 phage.

Francis Crick

Rosalind Franklin

Raymond Gosling

In 1953 James D. Watson and Francis Crick suggested what is now accepted as the first correct double-helix model of DNA structure in the journal Nature. Their double-helix, molecular model of DNA was then based on a single X-ray diffraction image taken by Rosalind Franklin and Raymond Gosling in May 1952, as well as the information that the DNA bases were paired, that is also obtained through private communications from Erwin Chargaff in the previous years. Chargaff's rules played a very important role in establishing double-helix configurations for B-DNA as well as A-DNA. Experimental evidence supporting the Watson and Crick model were published in a series of five articles in the same issue of Nature. Of these, Franklin and Gosling's paper was the first publication of their own X-ray diffraction data and original analysis method that partially supported the Watson and Crick model; this issue also contained an article on DNA structure by Maurice Wilkins and two of his colleagues, whose analysis and in vivo B-DNA X-ray patterns also supported the presence in vivo of the double-helical DNA configurations as proposed by Crick and Watson for their double-helix molecular model of DNA in the previous two pages of Nature. In 1962, after Franklin's death, Watson, Crick, and Wilkins jointly received the Nobel Prize in Physiology or Medicine. Unfortunately, Nobel rules of the time allowed only living recipients, but a vigorous debate continue on who should receive credit for the discovery. In an influential presentation in 1957, Crick laid out the "Central Dogma" of molecular biology, which foretold the relationship between DNA, RNA, and proteins, and articulated the "adaptor hypothesis". Final confirmation of the replication mechanism that was implied by the double-helical structure followed in 1958 through the Meselson-Stahl experiment. Further work by Crick and co-workers showed that the genetic code was based on non-overlapping triplets of bases, called codons, allowing Har Gobind Khorana, Robert W. Holley and Marshall Warren Nirenberg to decipher the genetic code. These findings represent the birth of molecular biology.

DNA

Deoxyribonucleic Acid (DNA) is a nucleic acid which carries genetic instructions for the biological development of all cellular forms of life and many viruses. DNA is sometimes referred to as the molecule of heredity as it is inherited and used to propagate traits. During reproduction, it is replicated and transmitted to offspring. In bacteria and other simple cell organisms, DNA is distributed more or less throughout the cell. In the complex cells that make up plants, animals and in other multi-celled organisms, most of the DNA is found in the chromosomes, which are located in the cell nucleus. The energy generating organelles known as chloroplasts and mitochondria also carry DNA.

Watson and Crick determine that nucleotides are join together to form long strands of DNA and each molecule consist of two strand that join together and wrap around each other to form a double helix(Fig. 2.52). Each chain of long polynucleotide composed of four types of nucleotides sub units. Hydrogen bonds between the base portions of the nucleotides hold the two chains together. In the case of the nucleotides in the DNA, the sugar is deoxyribose attaches to a single phosphate group, and the base may be either adenine, cytosine, guanine or thymine. The nucleotides are covalently linked together in a chain through the sugars and phosphates, which thus form a back bone of alternating sugar-phosphate- sugar-phosphate. The way in which the nucleotide subunits are linked together gives a DNA strand a chemical polarity (Fig. 2.53). This polarity in a DNA chain is indicated by referring to one end as the 3′ end and the other as the 5′ end. The three dimensional structure of DNA—the double helix—arises from the chemical and structural features of its two polynucleotide chains. These two chains are held together by hydrogen bonding between the bases on the different strands, all the bases are on the inside of the double helix, and sugar-phosphate backbones are on the outside. In each case, a two-ring base (a purine) spared with a single-ring base (a pyrimidine); A always pairs with T, and G paired with C. This complimentary base pairing enables the base pair to be packed in the energetically most favourable arrangement in the interior of the double helix. The two types of base pairs form different numbers of hydrogen bonds, AT forming two hydrogen bonds, and GC forming three hydrogen bonds (Fig. 2.54). DNA with high GC-content is more stable than DNA with low GC-content, but contrary to popular belief, this is not due to the extra hydrogen bond of a GC base pair but rather the contribution of stacking interactions (hydrogen bonding merely provides specificity of the pairing, not stability). As a result, it is both the percentage of GC base pairs and the overall length of a DNA double helix that determine the strength of the association between the two strands of DNA. Long DNA helices with a high GC content have stronger-interacting strands, while short helices with high AT content have weaker-interacting strands. A consequence of these base pairing requirements is that each strand of a DNA molecule contain a sequence of nucleotides that is exactly complimentary to the nucleotide sequence of its partner strands. The two strands of nucleotides in a double helix are considered antiparallel because the polarity of each strands are reversed relative to each other.

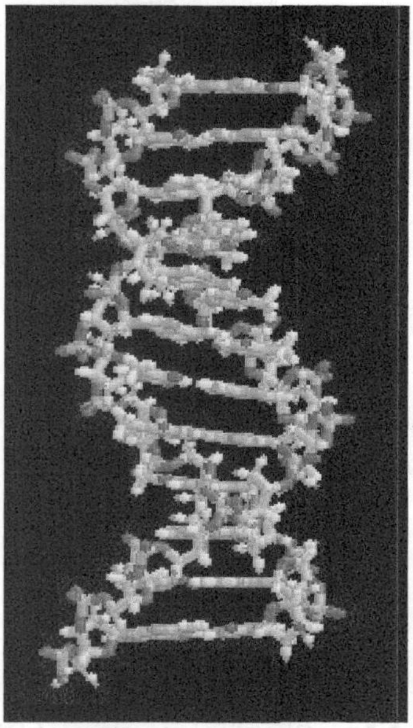

Fig. 2.52 The structure of DNA double helix

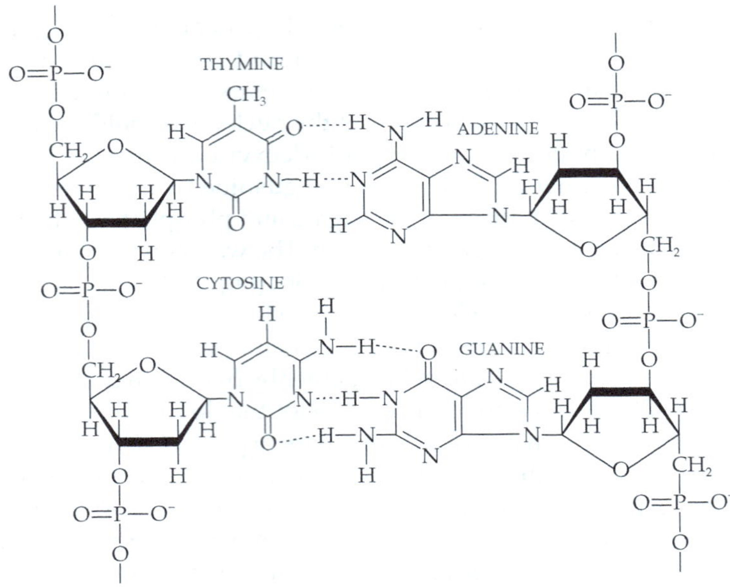

Chemical structure of DNA (A)

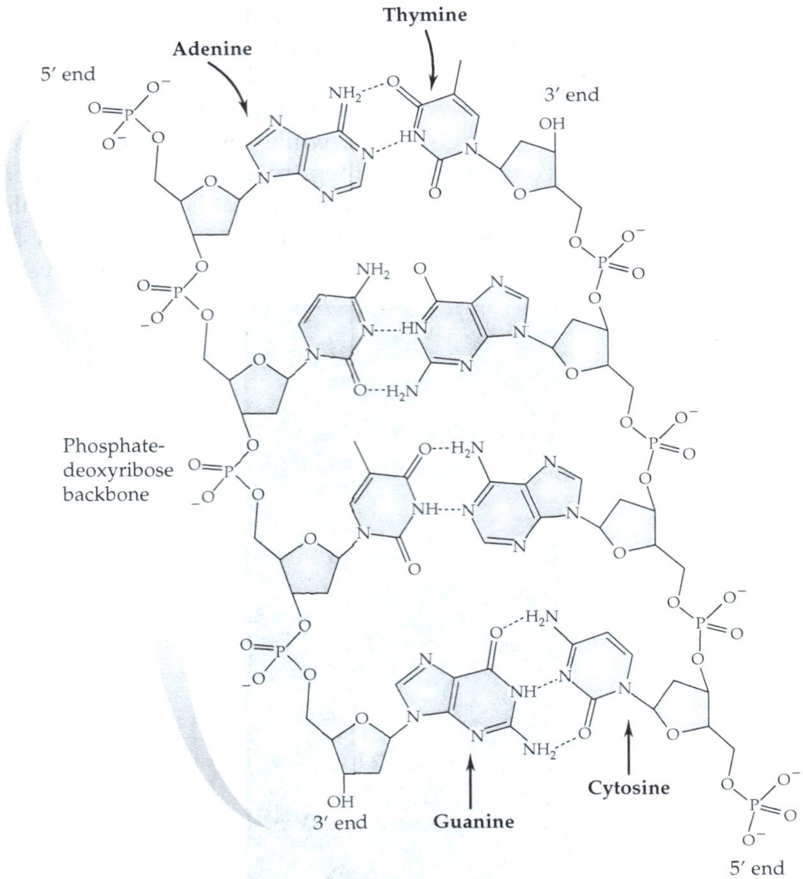

Fig. 2.53 Chemical structure of DNA

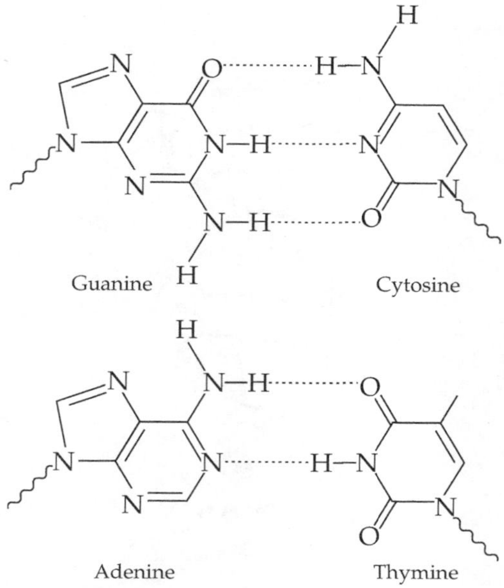

Fig. 2.54 Hydrogen bonds between the pairs are shown as dotted lines

Deferent Forms of DNA Double Helix

Double stranded DNA is a confirmationally variable molecule. In the following section we discussed it major conformational states.

B-DNA: B-DNA is regarded as the native form because its X-ray pattern resembles that of the DNA in intact sperm heads. It consists of two polynucleotide strands that wind about a common axis with a right handed twist to form an ~ 20-Å-diameter double helix. The two strands are antiparallel and wrap around each other such that they cannot be separated without unbinding the helix (a phenomenon known as plectonemic coiling). The bases occupy the core of the helix and sugar-phosphate chains are coiled about its periphery (Fig. 2.55). The planes of the bases are nearly perpendicular to the helix axis. Each base is hydrogen bonded to a base on the opposite strand to form a planer base pair. The ideal B-DNA helix has 10 base pairs per turn and since the aromatic bases have van der Waals thickness of 3.4 Å and are partially stacked on each other. B-DNA has two deep exterior grooves that wind between its sugar-phosphate chains as a consequence of the helix axis passing through the approximate centre of each base pair. The minor groove is that in which the C1′-helix axis- C1′ angle is $<180°$, where as the major groove opens towards the opposite edge of each base pair.

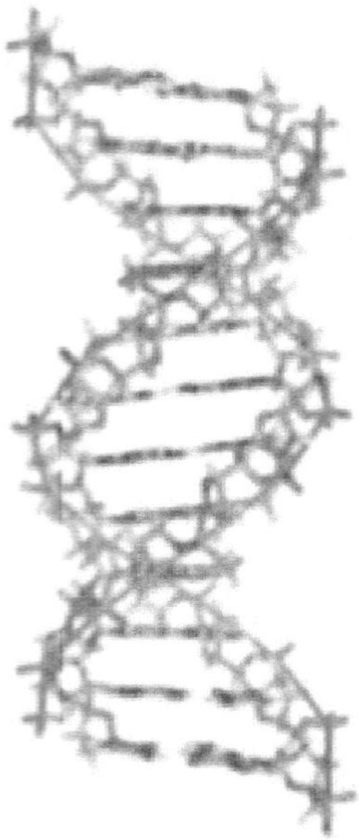

Fig. 2.55 The structure of B-DNA

***A-DNA*:** B-DNA undergoes a reversible conformational change, when the relative humidity is reduced 75%, called A-DNA. Fibre X-ray studies indicate that A-DNA forms a wider and flatter right-handed helix than does B-DNA. A-DNA has 11 bp per turn and a pitch of 28Å, which gives a A-DNAan axial hole. A-DNA's most striking feature is that the planes of its base pairs are tilted20°, with respect to the helix axis. Since, its helix axis does not pass through its base pairs; A-DNA has a deep major groove and a very shallow minor groove (Fig. 2.56).

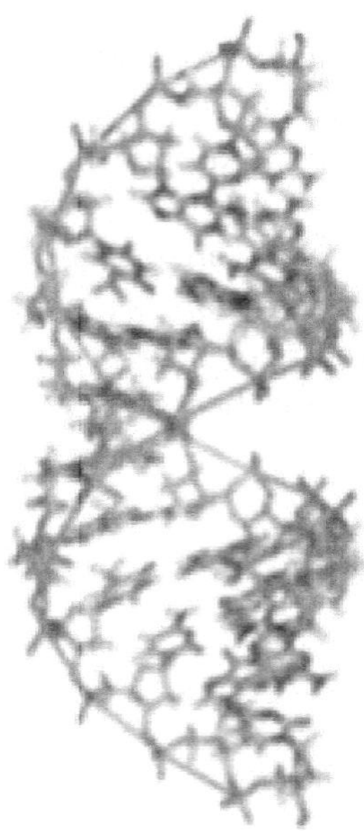

Fig. 2.56 The structure of A-DNA

Z-DNA: The Z-DNA confirmation is most readily assumed by DNA segments with alternating purine-pyrimidine base sequence. In Z-DNA 12 base pairs per helical turn, with a rise of 0.38 nm per base pairs, a pitch of 45Å held. It contains a deep minor groove and no major groove. Z-DNA therefore resembles a left handed drill bit in appearance. The base pair in Z-DNA are clipped 180° relative to those in B-DNA. Z-DNA follows a zigzag path around the helix (Fig. 2.57).

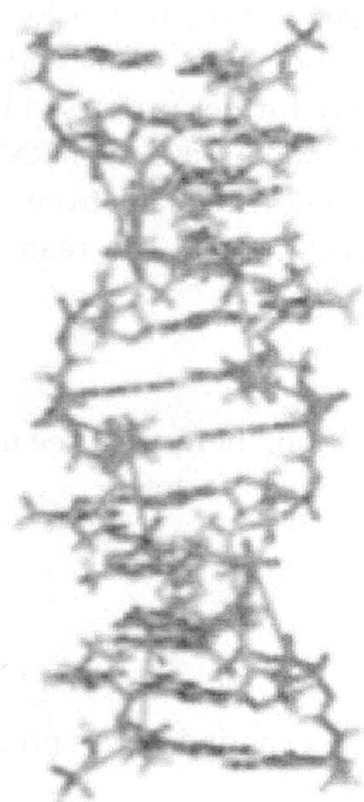

Fig. 2.57 The structure of Z-DNA

DNA Provides a Mechanism for Heredity

Genetic information is carried in the linear sequence of nucleotides in DNA. With the process of gene expression, cell translates the nucleotide sequence of a gene in to the amino acid sequence of a protein. Fig. 2.58, the complete set of information in an organism's DNA is called its genome and it carries the information for all the proteins the organism will ever synthesize.

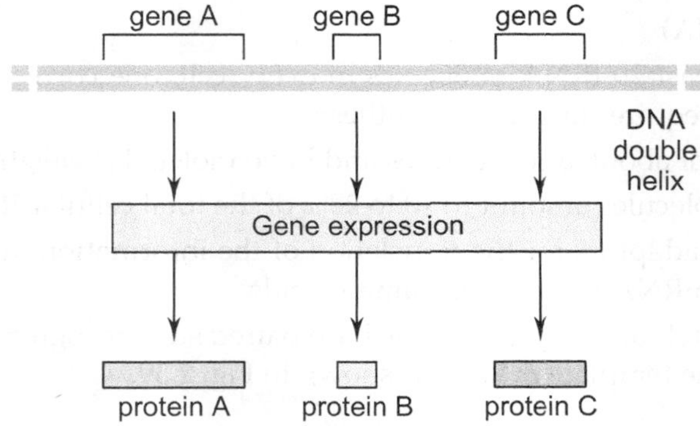

Fig. 2.58 The mechanism of heredity

Structure of RNA

Ribonucleic acid (RNA) is a biologically significant type of molecule that consists of a long chain of nucleotide units. Each nucleotide consists of a nitrogenous base, a ribose sugar, and a phosphate. RNA is very similar to DNA, but differs in a few important structural details: in the cell, RNA is usually single-stranded, while DNA is usually double-stranded; RNA nucleotides contain ribose while DNA contains deoxyribose (a type of ribose that lacks one oxygen atom); and RNA has the base uracil rather than thymine that is present in DNA. RNA is transcribed from DNA by enzymes called RNA polymerases and is generally further processed by other enzymes. RNA is central to protein synthesis. Here, a type of RNA called messenger RNA carries information from DNA to structures called ribosomes. These ribosomes are made from proteins and ribosomal RNAs, which come together to form a molecular machine that can read messenger RNAs and translate the information they carry into proteins.

Types of RNA

On the bases of their structural organization, there are three main cases of RNA molecules exist in all prokaryotic and eukaryotic organisms. These are:

(*a*) Messenger RNA (mRNA)
(*b*) Transfer RNA (tRNA)
(*c*) Ribosomal RNA (rRNA)

(*a*) **Messenger RNA (mRNA)**

1. The Messenger RNA is single stranded RNA and complimentary to the sense strand of DNA.
2. It is a most heterogeneous in size and stability.
3. The 5′ terminus of mRNA is capped by a 7-methylguanosine triphosphate which linked to an 2′-O methylribonucleoside at its 5′- hydroxyl through the 3-phosphate. The protein synthesizing machinery begins translating the mRNA in to proteins at the 5′ or capped terminus. The 3′ hydroxyl terminus has attached a polymer of adenylate residues 20–250 nucleotide in length.
4. It has a large molecular Wight of 3,000 to 50,000 to have the coded information corresponding to long polypeptide chain.

(*b*) **Transfer RNA (tRNA)**

1. There are atleast 20 tRNA molecules in every cell, one corresponds to each of the 20 amino acids require for protein synthesis.
2. They consist of about75 nucleotides and have molecular weight of 25,000.
3. The tRNA molecules amount to 10 to 20% of the total cellular RNA molecules.
4. This serve as adaptors for the translation of the information in the sequence of nucleotides of the mRNA in to specific amino acids.
5. The anti codon loop at the end of the base paired stem recognizes the triplet nucleotide or codon of the template mRNA as shown in Fig. 2.59.

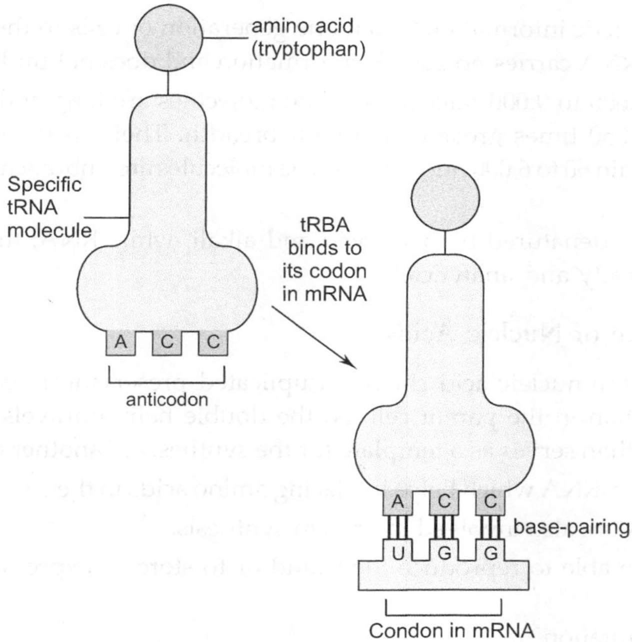

Fig. 2.59 Structure of tRNA

(*c*) **Ribosomal RNA (rRNA)**

1. Ribosomes are nucleoprotein particles and reticular granules are 100–150Å, in diameter which acts as the machinery for the synthesis of protein from mRNA templates.
2. On the ribosomes, the mRNA and tRNA molecules interact to translate in to specific protein molecule the information transcribed from the gene.
3. Ribosomal particles are very complex. The mammalian ribosome contain two major nucleoprotein subunits—a larger one (60S) and a smaller one (40S).
4. The (60S) subunit contains a 5S ribosomal RNA, a 5.8 S rRNA and 28S rRNA.
5. The smaller 40S subunit contains a single 18S rRNA and approximately 30 polypeptide chains.
6. In the cytoplasm, the ribosomes remain quite stable and capable of many translations.

Comparison between DNA and RNA

RNA and DNA are both nucleic acids, but differ in three main ways.

1. DNA, which is double-stranded, RNA is a single-stranded molecule in most of its biological roles and has a much shorter chain of nucleotides.
2. DNA contains deoxyribose, RNA contains ribose (there is no hydroxyl group attached to the pentose ring in the 2′ position in DNA). These hydroxyl groups make RNA less stable than DNA because it is more prone to hydrolysis.
3. The complementary base to adenine is not thymine, as it is in DNA, but rather uracil, which is an unmethylated form of thymine present in RNA.
4. DNA is present in nucleus, probably in the chromosomes. While RNA is present in cytoplasm and nucleolus.

5. DNA carries "genetic information" from one generation of cells to the next and undergoes mutation while RNA carries no genetic information and does not undergo mutation.
6. DNA contains 1,600 to 9.000 nucleotides. The molecules are long and thread like having a length of about 250 times greater than their breadth. Their structure is highly complex. Whilst RNA contain 60 to 6,000 nucleotides. The molecules are unbranched and there structure is less rigid.
7. DNA molecule is denatured by heat, acid and alkali, while RNA; the alkali hydrolysis is useful diagnostically and analytically.

Biological Significance of Nucleic Acids

1. In cell division, the nucleic acid chain is duplicated preserving in each daughter cell the information contain in the parent cell. So the double helix unravels and each of the two original strands than serves as a template for the synthesis of another complimentary chain.
2. DNA produces a mRNA which helps in placing amino acids in the code for protein synthesis.
3. rRNA and tRNA are also involved in protein synthesis.
4. Nucleic acids are able to reproduce their kind or to store or express and transmit genetic information.
5. They undergo mutation.
6. According to the pairing rule in DNA, adenine can only combine with thymine and guanine only with cytosine. The newly synthesized strand will be exactly constituted in its nucleotide sequence as was the original complimentary strand of the parent strand. The result is the synthesis of two pairs of strands.
7. RNA functions primarily in the cytoplasm of the cell as a template in connection with the synthesis of protein.
8. Adenylic acid in combination with two molecules of phosphate is the biochemical unit of energy exchange in all cells which is said to be ATP.
9. Biological oxidation-reduction involves the transport of hydrogen atoms or electrons through organized systems of substances called hydrogen acceptors or electron transport agents. The hydrogen acceptors are nucleotides such as NAD, FAD etc.

2.3 WHAT ARE ENZYMES?

Enzymes are the most remarkable and highly specialized proteins. Enzymes have extraordinary catalytic power, often far greater than that of synthetic or inorganic catalyst. They have a high degree of specificity for their substrate, they accelerate chemical reactions immensely, and they function in aqueous solutions under very mild conditions of temperature and pH. They are soluble in nature, colloidal organic catalysts formed by living cells, inactive at 0°C and destroyed by moist heat at 100°C.

Enzymes may be defined as biocatalysts synthesized by living cells. They are protein in nature (exception- RNA acting as ribozyme), colloidal and thermolabile in character, and specific in their action.

Enzymes are central to every biochemical process. The mechanism of enzyme catalysis is similar in principle to other types of chemical catalysis. By providing an alternative reaction route and by stabilizing intermediates the enzyme reduces the energy required to reach the highest energy transition state of the reaction. The reduction of activation energy (Ea) increases the number of reactant molecules with enough energy to reach the activation energy and form the product.

Actually, there are two possible important theories for enzyme reaction takes place; first is collision theory and second is transition state theory.

Collision theory proposes that reactions take place by the collision of the reactant molecules. That is, more is the concentration of the reactants, more are the chances for the reactants to collide, and hence more will be the rate of reaction. But, it is essential that all collision may not necessarily result in the reaction to proceed to produce products.

Transition state theory proposes that the collision of certain molecules, which have crossed certain potential energy barrier, alone will result in the reaction to take place. This potential energy barrier is known as activation energy (Fig. 2.60). The bio catalysis of the enzymatic reactions can be explained by the concept that the enzymes form with the substrate an intermediate complex which further decomposes to form the bio products Fig. 2.61). For the reaction

$$S \rightarrow P$$

Where, S is the substrate and P is the product, the transition state proposes a halfway point and unstable intermediate complex, where the bonds of S are distorted sufficiently so that conversion of P becomes possible. The rate of reaction obviously depends on the number of molecules of S that enter the transition state per unit time. Therefore, the rate of reaction can be increased by raising the temperature or lowering the activation energy. The enzyme complexes would selectively lower the energies of activation of vital biochemical reactions. Therefore, the enzymes can carry out the reactions at ambient temperatures. This makes it possible for an enzyme catalysed reaction at 25°C to proceed at 10^6–10^{15} times faster as compared to an uncatalyzed reaction.

Generally Enzymes are Proteins

With the omission of a small group of catalytic RNA molecules, all enzymes are proteins. Their catalytic activity depends on the integrity of their native protein conformation. If an enzyme is denatured or dissociated in to subunits, catalytic activity is usually lost. If an enzyme is broken down in to its component amino acids, its catalytic activity is always destroyed. Thus the primary, secondary, tertiary and quaternary structures of protein enzymes are essential to their catalytic activity. The molecular weights of enzymes like other proteins have ranging from about 12,000 to over 1 million. Some enzymes require no chemical groups for activity other than their amino acid residues. Other require an additional chemical component called a **cofactor**—either one or more inorganic ions, such as Fe^{++}, Mg^{++}, Mn^{++}, or Zn^{++} (Table 2.18), or a complex organic or metalloorganic molecule called a coenzyme. Coenzymes are heat-stable; many reactions of substrates are catalyzed by enzymes only in the presence of a specific non-protein organic molecule called the **coenzyme** (Table 2.19).

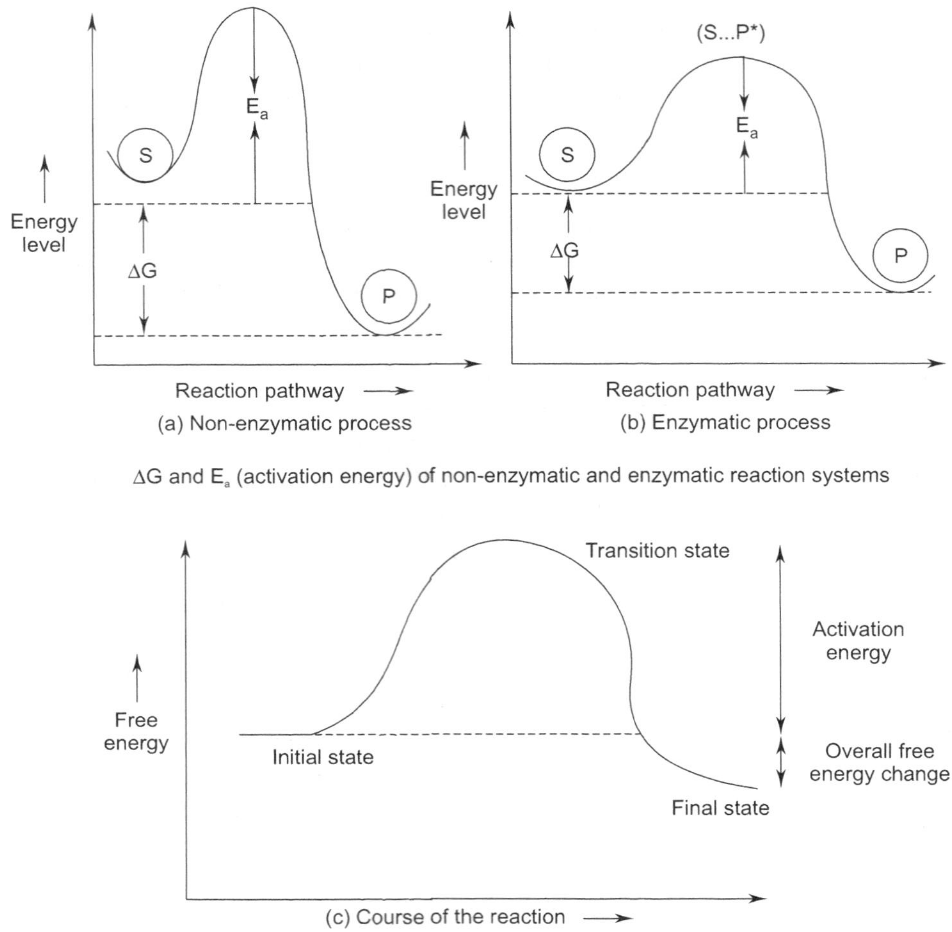

Fig. 2.60 Free energy change for a reaction taking place via formation of a transition state

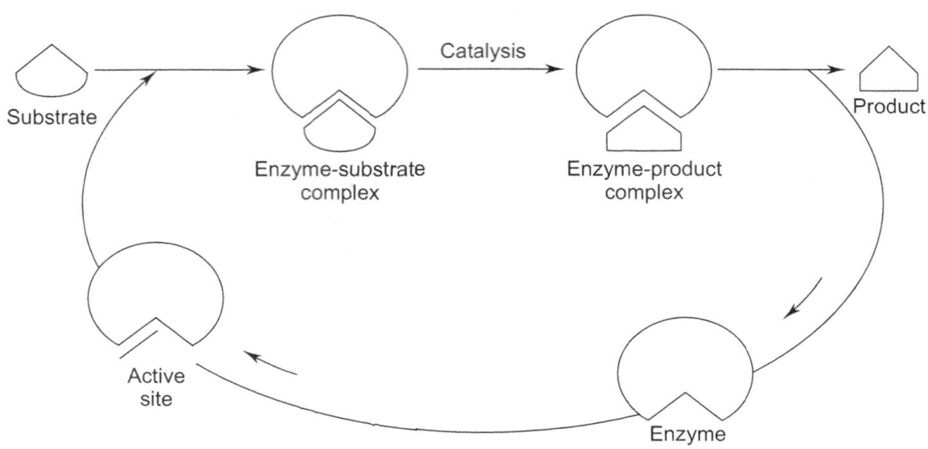

Fig. 2.61 Conversion of substrate to product by enzymatic process

Table 2.18 Various inorganic elements that provide as cofactor for enzymes

Cu^{++}	Cytochrome oxidase
Fe^{++} or Fe^{+++}	Cytochrome oxidase, Catalase, Peroxidase
K^{+}	Pyruvate kinase
Mg^{++}	Hexokinase, Glucose 6-phosphate, Pyruvate kinase
Mn^{++}	Arginase, Ribonucleotide reductase
Mo	Dinitrogenase
Ni^{++}	Urease
Se	Glutathione peroxidase
Zn^{++}	Carbonic anhydrase, Alcohol dehydrogenase, Carboxypeptidases A and B

Some enzymes require both a coenzyme and one or more metal ions for activity. A coenzyme or metal ion that is very tightly or even covalently bound to the enzyme protein is called a **prosthetic group**. Coenzymes combine with the **apoenzyme** or **apoprotein** (the protein part) to form **holoenzyme**. Coenzymes function as transient carriers of specific functional groups. They are often derived from vitamins, organic nutrients required in small amount in diet.

Table 2.19 Various coenzymes that provide as transient carriers of specific atoms or functional groups

Coenzyme	Example of chemical groups transferred	Dietary precursor in mammals
Biocytin	CO_2	Biotin
Coenzyme A	Acyl groups	Pantothenic acid and other compounds
5′-Deoxyadenosylcobalamin (coenzyme B_{12})	H atoms and alkyl groups	Vitamin B_{12}
Flavin adenine dinucleotide	Electrons	Riboflavin (vitamin B_2)
Lipoate	Electrons and acyl groups	Not required in diet
Nicotinamide adenine dinucleotide	Hydride ion (: H^-)	Nicotinic acid (niacin)
Pyridoxal phosphate	Amino group	Pyridoxine (vitamin B_6)
Tetrahydrofolate	One-carbon groups	folate
Thiamine pyrophosphate	Aldehydes	Thiamine (vitamin B_1)

2.3.1 Classification and Nomenclature of Enzymes

Based on catalyzed reactions, the nomenclature committee of the International Union of Biochemistry and Molecular Biology (IUBMB) recommended following 6 major classes of the enzyme.

1. **Oxidoreductases:** Enzyme catalyzes a variety of oxidation-reduction reactions. Common names include dehydrogenase, oxidase, reductase and catalase.
2. **Transferases:** They catalyze the transfers of some groups (acetyl, methyl, phosphate, etc.). Common names include acetyltransferase, methylase, protein kinase and polymerase. The first three subclasses play major roles in the regulation of cellular processes. Their chemical reactions are shown in Fig. 2.62. The polymerase is essential for the synthesis of DNA and RNA.

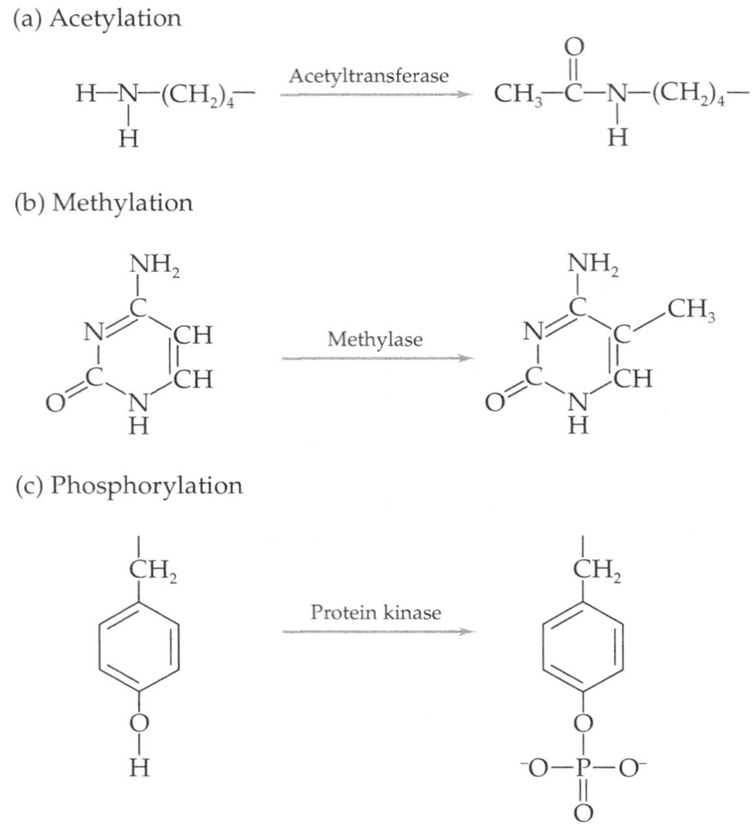

Fig. 2.62 Three major regulatory chemical reactions. (*a*) Acetylation-addition of an acetyl group to lysine's R group by acetyltransferase. (*b*) Methylation-addition of a methyl group to DNA's base (*e.g.*, cytosine) by methylase. (*c*) Phosphorylation-addition of a phosphate group to the R group of tyrosine, serine or threonine (only tyrosine is shown here) by protein kinase

3. **Hydrolases:** Catalyze hydrolysis reactions where a molecule is split into two or more smaller molecules by the addition of water. Common examples are given below.

   Proteases: Splits protein molecules. Examples: HIV protease and caspase. HIV protease is essential for HIV replication. Caspase plays a major role in apoptosis.

   Nucleases: Splits nucleic acids (DNA and RNA). Based on the substrate type, they are divided into RNAs and DNAs. RNAs catalyzes the hydrolysis of RNA and DNAs acts on DNA. They may also be divided into exonuclease and endonuclease. The exonuclease progressively splits off single nucleotides from one end of DNA or RNA. The endonuclease splits DNA or RNA at internal sites.

   Phosphatase: Catalyzes dephosphorylation (removal of phosphate groups). Example: calcineurin. The immunosuppressive drugs and Cyclosporin A are the inhibitors of calcineurin.
4. **Lyases:** Enzymes specialized in the addition or removal of water, ammonia and CO_2. It catalyzes the cleavage of C-C, C-O, C-S and C-N bonds by means other than hydrolysis or oxidation. Common names include decarboxylase and aldolase.
5. **Isomerases:** Catalyze atomic rearrangements within a molecule. It involved in all the isomerisation reactions. Examples include (Fig. 2.63) rotamase, protein disulfide isomerase (PDI), epimerase and racemase.

6. **Ligases:** catalyze the reaction which joins two molecules. Examples include peptide synthase, aminoacyl-tRNA synthetase, DNA ligase and RNA ligase.

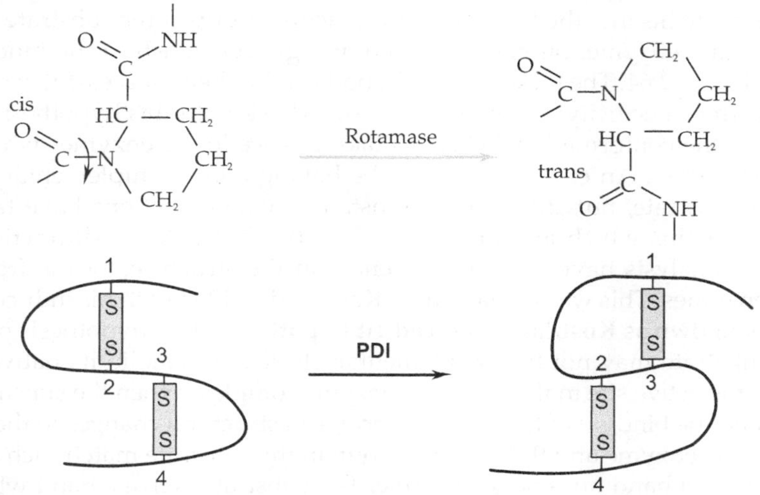

Fig. 2.63 The role of rotamase and protein disulfide isomerase (PDI). The reactions catalyzed by the two enzymes can assist a peptide chain to fold into a correct three-dimensional structure

[The word ***OTHLIL*** (first letter in each class) may be memorised to keep in mind the six classes of enzymes in the correct order].

Each class in turn is subdivided into many sub-classes which are further divided. A four digit Enzyme Commission (E.C.) number is assigned to each enzyme representing the class (first digit), sub-class second digit, sub-sub class (third digit) and the individual enzyme (fourth digit). Each enzyme is given a specific name indicating the substrate, coenzyme (if any) and the type of the reaction catalyzed by the enzyme. Although, the IUB names for the enzymes are specific and unambiguous, they have not been accepted for general use as they are complex and cumbersome to remember. Therefore, the trivial name, along with the E.C. numbers as and when needed, are commonly used and widely accepted.

2.3.2 Enzyme Specificity

Enzymes as biocatalysts are predominantly credited for their specificity in action. Every enzyme generally catalyses only one particular type of reaction by selectively lowering the activation energy of only one of the several possible biochemical/chemical reactions. Therefore, only stereospecific groups of the substrate alone undergo the desired reaction to form the desired product.

2.3.3 Enzyme Specificity Hypotheses

A number of hypotheses have been proposed from time to time to elucidate the enzyme specificity in catalytic activity and its ability to interact with the substrates. The some well-known hypotheses are:

(*i*) Fischer lock-and-key hypothesis
(*ii*) Koshland induced fit hypothesis

Fischer lock-and-key hypothesis is probably the first theory proposed in 1890 by Fischer who conceived the concept of 'complementary structural features' between the enzyme and the substrate. The catalytic process is brought about because the substrate fits into the complementary site on the enzyme just as a key fits into the lock. Thus, the reacting group of the substrate gets struck with the catalytic site of the enzyme. Similarly, the binding groups attach to the binding sites in the enzyme as shown in Fig. 2.64. The lock-and-key hypothesis has been successful in explaining many features of the enzyme specificity; but one serious drawback with this hypothesis is that it could not explain some of the conformational changes taking place in the enzymes when they come in contact with the substrate. An enzyme may not be having a just complementary feature that is compatible to the substrate, but still there are instances when reactions have taken place; predominantly with proteins which are known for their flexibility. X-ray diffraction analysis and some spectroscopic analysis have shown differences in the structures of the free enzymes and substrate bound enzymes. This was explained by Koshland in 1958 with his induced fit hypothesis which is popularly known as **Koshland induced-fit hypothesis**. This hypothesis proposes that the structure of the substrate may not be complementary to the enzyme in its native form, but it is complementary to the active site in the substrate-enzyme complex. When the substrate and enzyme are brought together, the binding of the substrate brings in structural changes in the enzyme so that the catalytic site in the enzyme and the reacting group in the substrate match each other (Fig. 2.65). This is explained with a hand-and-glove example. The substrate is like a hand which is rigid and does not change its shape, but the enzyme is like a woollen glove which can be wrapped around the hand, and just as it (woollen glove) adjusts itself to wrap the hand, so does the enzyme. The reaction, which is catalyzed by yeast Hexokinase, is a typical example of induced fit hypothesis.

$$\text{D-hexose} + \text{ATP} \rightleftharpoons \text{D-hexose-6-phosphate} + \text{ATP}$$

2.3.4 Properties of Enzymes

1. Enzymes are usually very specific as to which reactions they catalyze and the substrates that are involved in these reactions. Complementary shape, charge and hydrophilic/hydrophobic characteristics of enzymes and substrates are responsible for this specificity.
2. Enzymes are proteins in nature. Subsequently the amino, carboxyl and sulfhydryl groups of the side chains of amino acids are available for linkage between polypeptide chains.
3. The above mentioned groups together with others such as the imidazole ring and the alcoholic group of serine are also responsible for the union between substrate and enzyme. These groups are termed "**active groups**" and the region of the protein surface at which they are located is termed "**active centre**".
4. The enzyme-substrate complex theory assumes combination of enzymes and substrate and then liberation of enzyme and the reaction product.

   Thus,

   $$E + S \rightleftharpoons [ES] \rightarrow E + P \text{ \{Reaction Product(s)\}}$$
5. From an energy stand point enzyme reactions are divided in to three classes:
   (*a*) Exergonic reaction means the system undergoes a loss of free energy, *e.g.*, Lipase, Catalase, Urease.
   (*b*) There is little change in free energy and the reactions come to equilibrium, so that product molecules and the substrate molecules are present in constant amount. Example: Glycogen + Inorganic phosphate = Glucose-1-phosphate, under the influence of phosphorylase.
   (*c*) Endergonic reaction means the supply of energy in order to proceed the reaction.

6. Many enzymes are conjugated proteins. The prosthetic groups are readily detached. These enzymes have special role in metabolic processes. The enzymes catalase and peroxidase have cofactors, but no coenzyme.

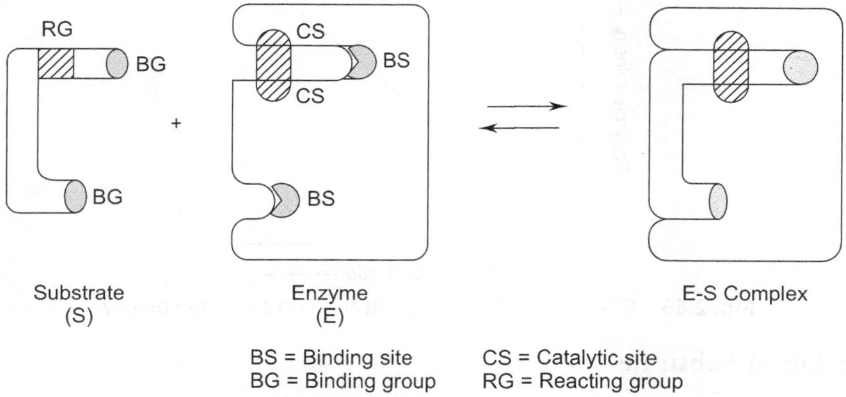

Fig. 2.64 lock-and-key hypothesis

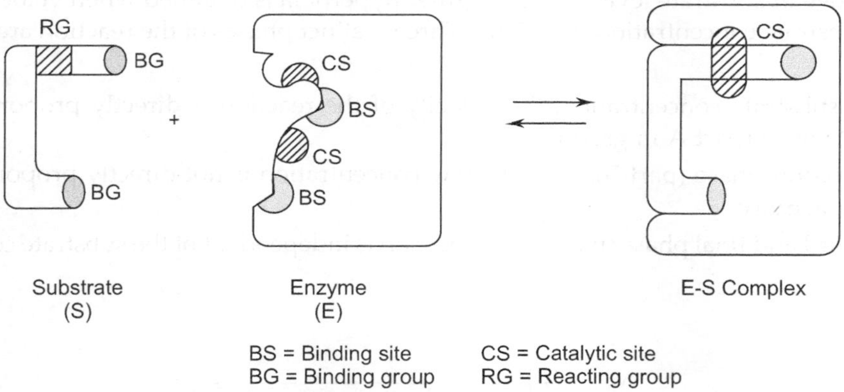

Fig. 2.65 Induced-fit model

2.3.5 Factors Affecting Enzyme Activity

The contact between the enzyme and substrate is the most essential pre-requisite for enzyme activity. The following important factors that influence the velocity of the enzyme reaction are discussed below:

A. Concentration of Enzyme

As the concentration of the enzyme is increased, the velocity of the reaction proportionately increases (Fig. 2.66) in fact; this property of f enzyme is made use in determining the serum enzymes for the diagnosis of diseases. By using a known volume of serum, and keeping all the other factors (substrate pH, temperature etc.) at the optimum level, the enzyme could be assayed in the laboratory. The relative concentration of the serum enzyme measured through its activity is used for the diagnosis of diseases.

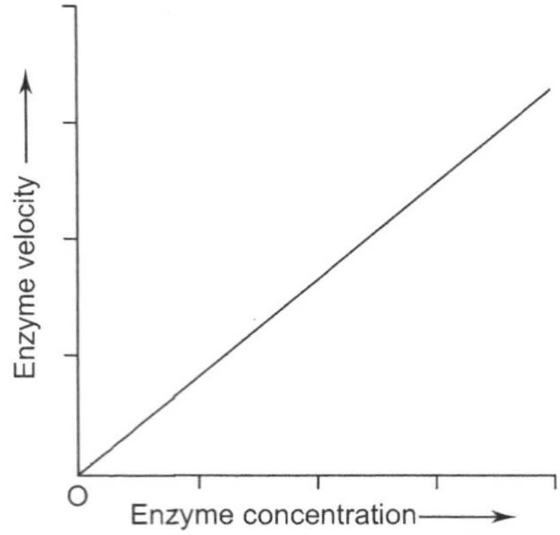

Fig. 2.66 Effect of enzyme concentration on enzyme velocity

B. Concentration of Substrate

Increase in the substrate concentration gradually increases the velocity of enzyme reaction within the limited range of substrate levels. A rectangular hyperbola is obtained when velocity is plotted against the substrate concentration (Fig. 2.67). Three distinct phases of the reaction are observed in the graph.

(*i*) At low substrate concentration, the velocity of the reaction is directly proportional to the substrate level (part A in graph).

(*ii*) In the second phase (part B), the substrate concentration is not directly proportional to the enzyme activity.

(*iii*) In the third and final phase (part C), the reaction is independent of the substrate concentration.

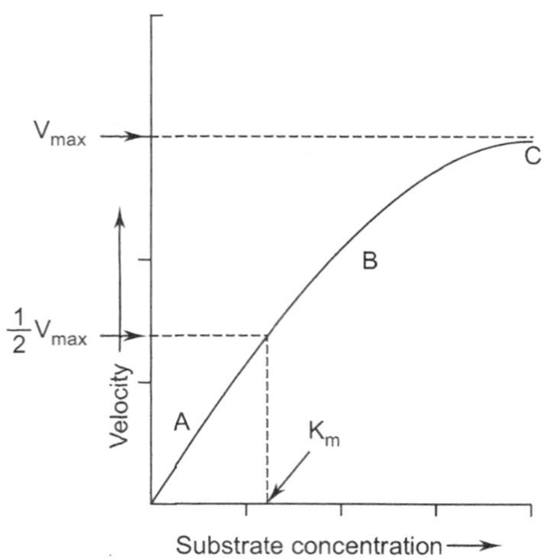

Fig. 2.67 Effect of substrate concentration on enzyme velocity (A-linear; B-curve; C-almost unchanged)

Enzyme Kinetics and K_m Value

The enzyme (E) and substrate (S) combine with each other to form an unstable enzyme- substrate complex (ES) for the formation of product (P).

$$E + S \underset{k_2}{\overset{k_1}{\rightleftharpoons}} ES \xrightarrow{k_3} E + P$$

Here k_1, k_2 and k_3 represent the velocity constant for the respective reactions, as indicated by arrows.

K_m, the Michaelis-Menten constant (or Brig's and Haldane's constant), is given by the formula

$$K_m = \frac{k_2 + k_3}{k_1}$$

$$v = \frac{V_{max}[S]}{K_m + [S]}$$

equation (1)

Where,

v = Measured velocity

V_{max} = Maximum velocity

S = Substrate concentration

K_m = Michaelis-Menten constant

Let us assume that the measured velocity (v) is equal to 1/2 V_{max}. Then, the equation (1) may be substituted as follows:

$$1/2\, V_{max} = \frac{V_{max}[S]}{K_m + [S]}$$

$$K_m + [S] = \frac{2V_{max}[S]}{V_{max}}$$

$$K_m + [S] = 2\,[S]$$

$$K_m = [S]$$

K_m or ***Michaelis-Menten constant*** is defined as the ***substrate concentration*** (expressed in moles/lit) to produce ***half-maximum velocity*** in an enzyme catalysed reaction. It indicates that half of the enzyme molecules (*i.e.*, 50%) are bound with the substrate molecules when the substrate concentration equals the K_m value. K_m, value is a constant and a characteristic feature of a given enzyme. It is a representative for measuring the strength of ES complex. A ***low K_m value indicates a strong affinity between enzyme and substrate***, where as a high K_m value reflects a weak affinity between them. For majority of enzymes, the K_m values are in the range of 10^{-5} to 10^{-2} moles.

Lineweaver-Burk Double Reciprocal Plot

For the determination of K_m value, the substrate saturation curve (Fig. 2.67) is not very accurate since V_{max} is approached asymptotically. By taking the reciprocals of the equation (1), a straight line graphic representation is obtained.

$$1/v = \frac{K_m + [S]}{V_{max}[S]}$$

$$1/v = \frac{K_m}{V_{max}} \times \frac{1}{[S]} + \frac{[S]}{V_{max}[S]}$$

$$1/v = \frac{K_m}{V_{max}} \times \frac{1}{[S]} + \frac{1}{V_{max}}$$

The above equation is similar to $y = ax + b$. Therefore, a plot of the reciprocal of the velocity $\{1/v\}$ vs. the reciprocal of the substrate concentration $\{1/[S]\}$ gives a straight line. Here, the slop is K_m/V_{max} and y intercept is $1/V_{max}$.

The Lineweaver-Burk plot is shown in Fig. 2.68. It is much easier to calculate the K_m From the intercept on x-axis which is $-(1/K_m)$. Further, the double reciprocal plot is useful in understanding the effect of various inhibitions.

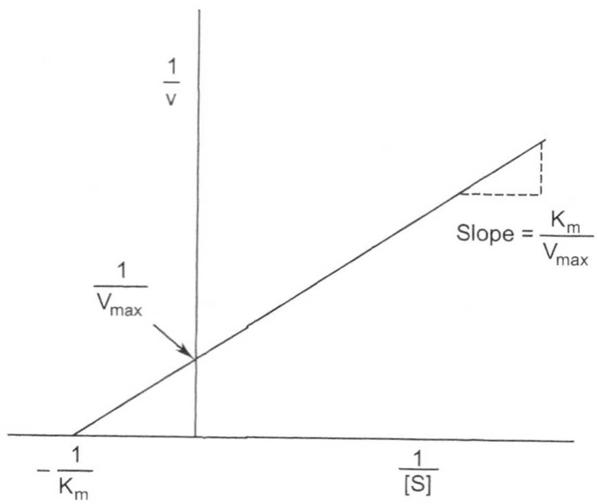

Fig. 2.68 Lineweaver-Burk double reciprocal plot

C. Effect of Temperature

Velocity of an enzyme reaction increases with increase in temperature up to a maximum and then declines. A bell-shaped curve is usually observed (Fig. 2.69)

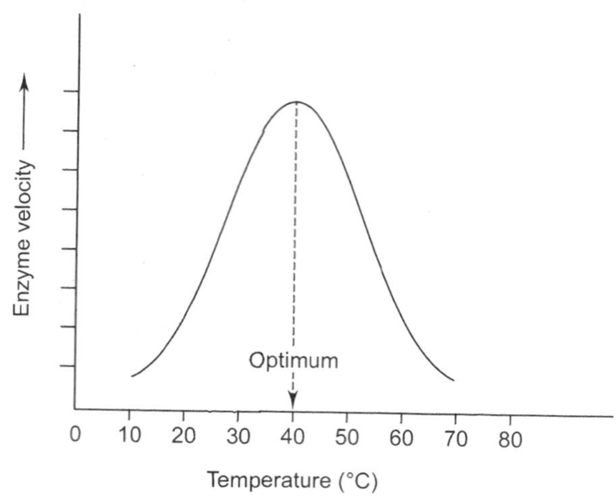

Fig. 2.69 Effect of temperature on enzyme velocity

Temperature coefficient or Q_{10} is defined as increase in enzyme velocity when the temperature is increased by 10°C. For a majority of enzyme, Q_{10} is 2 between 0°C and 40°C. Increase in temperature results in higher activation energy of the molecules and more molecular (enzyme and substrate) collision and interaction for the reaction to proceed faster.

The optimum temperature for most of the enzymes is between 40°C–45°C. However, a few enzymes (*e.g.*, venom phosphokinases, muscle adenylate kinase) are active even at 100°C. Some plant enzymes like urease have optimum activity around 60°C. This may be due to very stable structure and conformation of these enzymes.

In general, when the enzymes are exposed to a temperature above 50°C, **denaturation** leading to derangement in the native (tertiary) structure of the protein and active site are seen. Majority of the enzymes become inactive at higher temperature (above 70°C). It is worth noting here that the enzymes have been assigned optimal temperatures based on the laboratory work. These temperatures however, may have less relevance and biological significance in the living system.

D. Effect of pH

Increase in the hydrogen ion concentration (pH) considerably influences the enzyme activity and a bell-shaped curve is normally obtained (Fig. 2.70). Each enzyme has an optimum pH at which the velocity is maximum. Below and above this pH, the enzyme activity is much lower and at extreme pH, the enzyme-becomes totally inactive.

Most of the enzymes of higher organisms show optimum activity around neutral pH (6–8). There are, however, many exceptions like pepsin (1–2), acid phosphatase (4–5) and alkaline phosphatase (10–11). Enzymes from fungi and plants are most active in acidic pH (4–6). Hydrogen ions influence the enzyme activity by altering the ionic charges on the amino acids (particularly at the active site), substrate, ES complex etc.

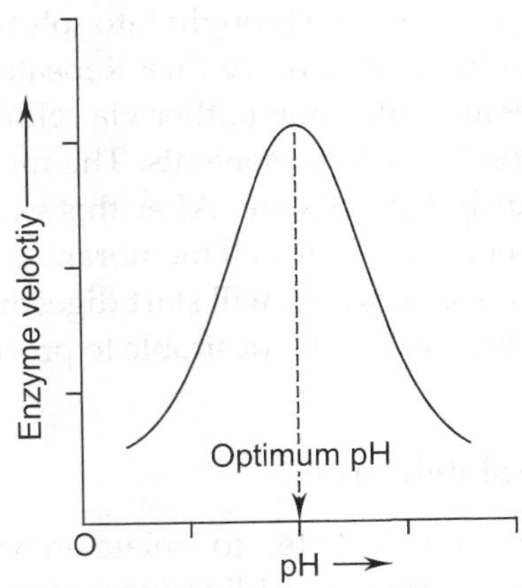

Fig. 2.70 Effect of pH on enzyme velocity

E. Effect of Light and Radiation

Exposure of enzymes to ultraviolet, beta, gamma, and X-rays inactivates certain enzymes due to the formation of peroxides, *e.g.*, UV rays inhibit salivary amylase activity.

2.3.6 Purification and Characterization of Enzymes

Presently, the use of enzymes in industrial sector is increasing due to the increase of industries, especially in food, beverage, textile, leather and paper industries. Beside it uses in industry, it can also be used in the treatment of industrial waste, such as cellulase which is able to convert cellulose of wood and paper wastes to ethanol. Due to the multidimensional functions and role of enzymes, it is important to develop a technique of enzyme isolation and purification in order to the least possible cost.

Methods of Enzyme Purification

The methods used in enzymes purification can roughly be divided into analytical and preparative methods. The distinction is not exact, but the deciding factor is the amount of enzymes that can practically be purified with that method. Analytical methods aim to detect and identify a protein in a mixture, whereas preparative methods aim to produce large quantities of the protein for other purposes, such as structural biology or industrial use. In general, the preparative methods can be used in analytical applications, but not the other way around. An analytical purification generally utilizes three properties to separate enzymes. First, enzymes may be purified according to their isoelectric points by running them through a pH graded gel or an ion exchange column. Second, enzymes can be separated according to their size or molecular weight via size exclusion chromatography or by SDS-PAGE (sodium dodecyl sulfate-polyacrylamide gel electrophoresis) analysis. Enzymes are often purified by using 2D-PAGE and are then analysed by peptide mass fingerprinting to establish the protein identity. This is very useful for scientific purposes and the detection limits for protein are nowadays very low and nanogram amounts of protein are sufficient for their analysis. Thirdly, enzymes may be separated by polarity/hydrophobicity via high performance liquid chromatography or reversed-phase chromatography.

Extraction

Depending on the source, the enzyme has to be brought into solution by breaking the tissue or cells containing it. There are several methods to achieve this: Repeated freezing and thawing, sonication, homogenization by high pressure, filtration (either via cellulose-based depth filters or cross-flow filtration), or permeabilization by organic solvents. The method of choice depends on how fragile the protein is and how sturdy the cells are. After this extraction process soluble proteins will be in the solvent, and can be separated from cell membranes, DNA etc., by centrifugation. The extraction process also extracts proteases, which will start digesting the proteins in the solution. If the protein is sensitive to proteolysis, it is usually desirable to proceed quickly, and keep the extract cooled, to slow down proteolysis.

Precipitation and Differential Solubilization

In bulk enzyme purification, a common first step to isolate enzyme is precipitation with ammonium sulfate $(NH_4)_2SO_4$. This is performed by adding increasing amounts of ammonium sulfate and collecting the different fractions of precipitate protein. One advantage of this method is that it can be performed inexpensively with very large volumes.

The first proteins to be purified are water-soluble proteins. Purification of integral membrane proteins requires disruption of the cell membrane in order to isolate any one particular protein from others that are in the same membrane compartment. Sometimes a particular membrane fraction can be isolated first, such as isolating mitochondria from cells before purifying a protein located in a mitochondrial membrane. A detergent such as Sodium Dodecyl Sulfate (SDS) can be used to dissolve cell membranes and keep membrane proteins in solution during purification; however, because SDS causes denaturation, milder detergents such as Triton X-100 or CHAPS can be used to retain the protein's native conformation during complete purification.

Ultracentrifugation

Centrifugation is a process that uses centrifugal force to separate mixtures of particles of varying masses or densities suspended in a liquid. When a vessel (typically a tube or bottle) containing a mixture of proteins or other particulate matter, such as bacterial cells, is rotated at high speeds, the angular momentum yields an outward force to each particle that is proportional to its mass. The tendency of a given particle to move through the liquid because of this force is offset by the resistance the liquid exerts on the particle. The net effect of "spinning" the sample in a centrifuge is that massive, small, and dense particles move outward faster than less massive particles or particles with more "drag" in the liquid. When suspensions of particles are "spun" in a centrifuge, a "pellet" may form at the bottom of the vessel that is enriched for the most massive particles with low drag in the liquid. Non-compacted particles still remaining mostly in the liquid are called the "supernatant" and can be removed from the vessel to separate the supernatant from the pellet. The rate of centrifugation is specified by the angular acceleration applied to the sample, typically measured in comparison to the *g*. If samples are centrifuged long enough, the particles in the vessel will reach equilibrium wherein the particles accumulate specifically at a point in the vessel where their buoyant density is balanced with centrifugal force. Such an "equilibrium" centrifugation can allow extensive purification of a given particle.

Sucrose gradient centrifugation: a linear concentration gradient of sugar (typically sucrose, glycerol, or a silica based density gradient media, like Percoll) is generated in a tube such that the highest concentration is on the bottom and lowest on top. Percoll is a trademark owned by GE Healthcare companies. A protein sample is then layered on top of the gradient and spun at high speeds in an ultracentrifuge. This causes heavy macromolecules to migrate towards the bottom of the tube faster than lighter material. During centrifugation in the absence of sucrose, as particles move farther and farther from the centre of rotation, they experience more and more centrifugal force (the further they move, the faster they move). The problem with this is that the useful separation range of within the vessel is restricted to a small observable window. Spinning a sample twice as long doesn't mean the particle of interest will go twice as far, in fact, it will go significantly further. However, when the proteins are moving through a sucrose gradient, they encounter liquid of increasing density and viscosity. A properly designed sucrose gradient will counteract the increasing centrifugal force so the particles move in close proportion to the time they have been in the centrifugal field. Samples separated by these gradients are referred to as "rate zonal" centrifugations. After separating the protein/particles, the gradient is then fractionated and collected.

Chromatographic Methods

Usually, an enzyme purification procedure contains one or more chromatographic steps. The basic procedure in chromatography is to flow the solution containing the protein through a column packed with various materials. Different proteins interact differently with the column material, and can thus be separated by the time required to pass the column, or the conditions required to elute the protein from the column. Usually, proteins are detected as they are coming off the column by their absorbance at 280 nm. Many different chromatographic methods exist:

Size Exclusion Chromatography/Gel Permeation Chromatography

Size Exclusion Chromatography (SEC) is also known as Gel Permeation Chromatography (GPC) or gel filtration chromatography and separates molecules according to their size (or more accurately according to their hydrodynamic diameter or hydrodynamic volume). Smaller molecules are able to enter the pores of the media and, therefore, molecules are trapped and removed from the flow of the mobile phase. The average residence time in the pores depends upon the effective size of the analyte molecules. However, molecules that are larger than the average pore size of the packing are excluded and thus suffer essentially no retension; such species are the first to be eluted. It is generally a low-resolution chromatography technique and thus it is often reserved for the final, "polishing" step of purification. It is also useful for determining the tertiary structure and quaternary structure of purified proteins, especially since it can be carried out under native solution conditions.

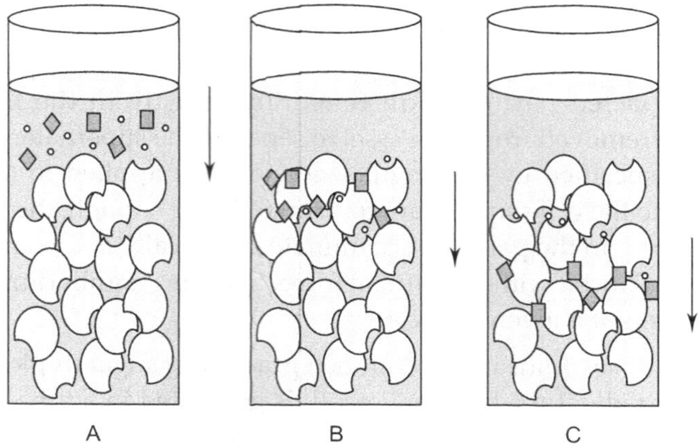

Fig. 2.71 (*a*) Size exclusion chromatography, **A:** Mixture of large molecules (diamonds) and small molecules in (circles) are applied to the top of a gel filtration column. **B:** Upon entering the column, the small molecules enter pores in the stationary phase matrix from which the large molecules are excluded. **C:** as the mobile phase flows down the column, the large, excluded molecules flow with it while the small molecules, which are temporarily sheltered from the flow when inside the pores, lag farther and farther behind

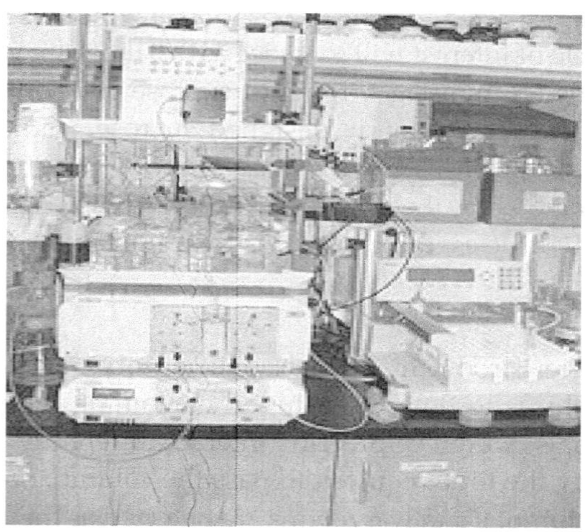

Fig. 2.71 (*b*) Chromatographic equipment, set up for a size exclusion chromatography

Ion Exchange Chromatography

Ion exchange chromatography separates compounds according to the nature and degree of their ionic charge. The column to be used is selected according to its type and strength of charge. Anion exchange resins have a positive charge and are used to retain and separate negatively charged compounds, while cation exchange resins have a negative charge and are used to separate positively charged molecules (Fig. 2.72).

Before the separation begins a buffer is pumped through the column to equilibrate the opposing charged ions. Upon injection of the sample, solute molecules will exchange with the buffer ions as each competes for the binding sites on the resin. The length of retention for each solute depends upon the strength of its charge. The most weakly charged compounds will elute first, followed by those with successively stronger charges. Because of the nature of the separating mechanism, pH, buffer type, buffer concentration, and temperature all play important roles in controlling the separation.

Ion exchange chromatography is a very powerful tool for use in protein purification and is frequently used in both analytical and preparative separations.

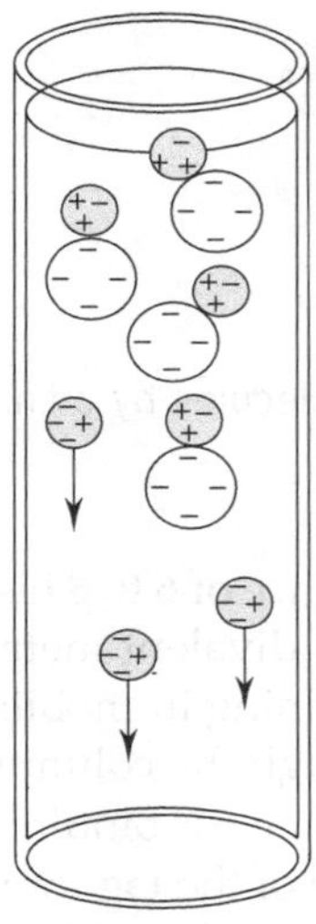

Fig. 2.72 Ion exchange chromatography separates proteins mainly according to their net charge

Affinity Chromatography

Affinity Chromatography is a separation technique based upon molecular conformation, which frequently utilizes application specific resins. These resins have ligands attached to their surfaces which are specific for the compounds to be separated. Most frequently, these ligands function in a fashion similar to that of antibody-antigen interactions. This "lock and key" fit between the ligand and its target compound makes it highly specific, frequently generating a single peak, while all else in the sample is unretained.

Many membrane proteins are glycoproteins and can be purified by lectin affinity chromatography. Detergent-solubilized proteins can be allowed to bind to a chromatography resin that has been modified to have a covalently attached lectin. Proteins that do not bind to the lectin are washed away and then specifically bound glycoproteins can be eluted by adding a high concentration of a sugar that competes with the bound glycoproteins at the lectin binding site. Some lectins have high affinity binding to oligosaccharides of glycoproteins that is hard to compete with sugars, and bound glycoproteins need to be released by denaturing the lectin.

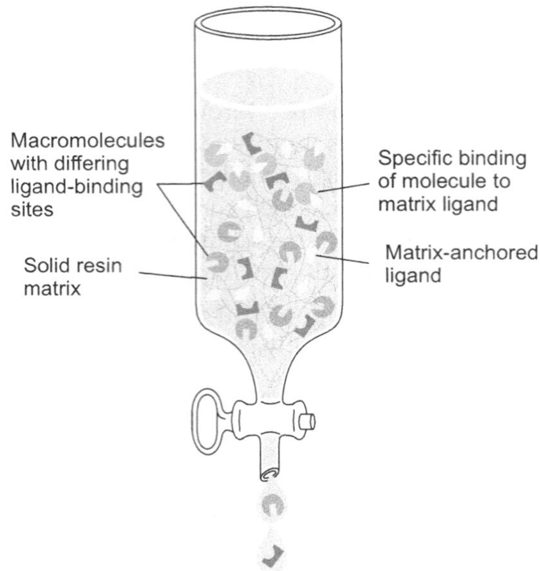

Fig. 2.73 The separation of macromolecules by affinity chromatography

Metal Binding

A common technique involves engineering a sequence of 6 to 8 histidines into the N- or C-terminal of the protein. The polyhistidine binds strongly to divalent metal ions such as nickel and cobalt. The protein can be passed through a column containing immobilized nickel ions, which binds the polyhistidine tag. All untagged proteins pass through the column. The protein can be eluted with imidazole, which competes with the polyhistidine tag for binding to the column, or by a decrease in pH (typically to 4.5), which decreases the affinity of the tag for the resin. While this procedure is generally used for the purification of recombinant proteins with an engineered affinity tag (such as a 6xHis tag or Clontech's HAT tag), it can also be used for natural proteins with an inherent affinity for divalent cations.

Immunoaffinity Chromatography

Immunoaffinity chromatography uses the specific binding of an antibody to the target protein to selectively purify the protein. The procedure involves immobilizing an antibody to a column material, which then selectively binds the protein, while everything else flows through. The protein can be eluted by changing the pH or the salinity. Because, this method does not involve engineering in a tag, it can be used for proteins from natural sources.

High Performance Liquid Chromatography

High performance liquid chromatography or high pressure liquid chromatography is a form of chromatography applying high pressure to drive the solutes through the column faster. This means that the diffusion is limited and the resolution is improved. The most common form is "reversed phase" HPLC, where the column material is hydrophobic. The proteins are eluted by a gradient of increasing amounts of an organic solvent, such as acetonitrile. The proteins elute according to their hydrophobicity. After purification by HPLC the protein is in a solution that only contains volatile compounds, and can easily be lyophilized. HPLC purification frequently results in denaturation of the purified proteins and is thus not applicable to proteins that do not spontaneously refold.

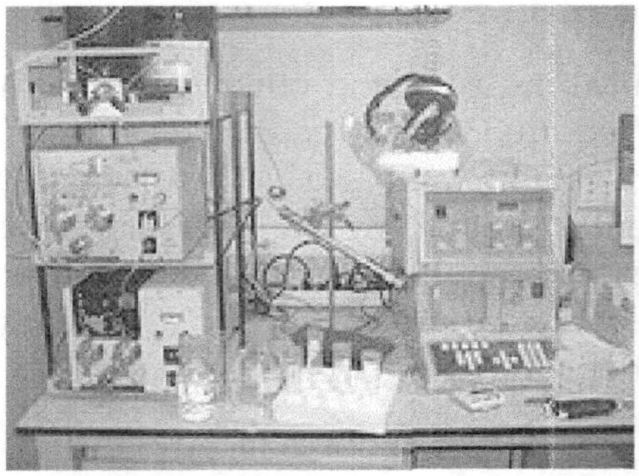

Fig. 2.74 A HPLC. From left to right: A pumping device generating a gradient of two different solvents, a steel enforced column and an apparatus for measuring the absorbance

Ultrafiltration

Ultrafiltration concentrates a protein solution using selective permeable membranes. The function of the membrane is to let the water and small molecules pass through while retaining the protein. The solution is forced against the membrane by mechanical pump or gas pressure or centrifugation. Membrane separation may be used to separate proteins after precipitation to remove many of the contaminants. Membranes can serve as a molecular sieve to separate solute molecules of different molecular size. Depending on molecular size cut off, different membrane can be used for the separation of different MW proteins. Ultra filters are used for macromolecules with a molecule weight range of 2000 to 500,000. Some ultra filtration membranes have anisotropic structure. In an anisotropic membrane, a thin skin with small pores is formed on top of a thick, highly porous structure. The thin layer provides selectivity, while the thicker layer provides mechanical supports.

Denaturing-Condition Electrophoresis

Gel electrophoresis is a common laboratory technique that can be used both as preparative and analytical method. The principle of electrophoresis relies on the movement of a charged ion in an electric field. In practice, the proteins are denatured in a solution containing a detergent (SDS). In these conditions, the proteins are unfolded and coated with negatively charged detergent molecules. The proteins in SDS-PAGE are separated on the sole basis of their size. In analytical methods, the proteins migrate as bands based on size. Each band can be detected using stains such as Coomassie blue dye or silver stain. Preparative methods to purify large amounts of protein require the extraction of the protein from the electrophoretic gel. This extraction may involve excision of the gel containing a band, or eluting the band directly off the gel as it runs off the end of the gel.

In the context of a purification strategy, denaturing condition electrophoresis provides an improved resolution over size exclusion chromatography, but does not scale to large quantity of proteins in a sample as well as the late chromatography columns.

An important non-denaturing electrophoretic procedure for isolating bioactive metalloproteins in complex protein mixtures is termed 'quantitative native continuous polyacrylamide gel electrophoresis (QPNC-PAGE).

Characterization of Purified Enzyme

The characterization of purified enzyme includes: determination of optimum pH, temperature, kinetic data, molecular weight, and stability toward temperature and pH.

2.3.7 Industrial Applications of Enzymes

Enzymes are used in the chemical industry and other industrial applications (Table 2.20) when extremely specific catalysts are required. However, enzymes in general are limited in the number of reactions they have evolved to catalyze and also by their lack of stability in organic solvents and at high temperatures. Consequently, protein engineering is an active area of research and involves attempts to create new enzymes with novel properties, either through rational design or in vitro evolution. These efforts have begun to be successful, and a few enzymes have now been designed "from scratch" to catalyse reactions that do not occur in nature.

Table 2.20

Application	Enzymes used	Uses
Food processing 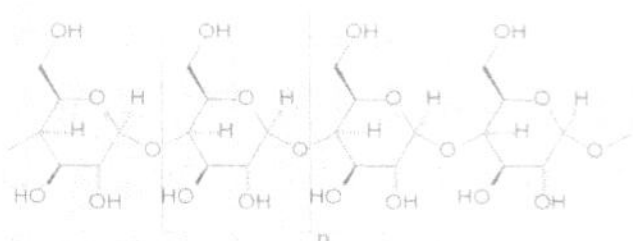 Amylases catalyze the release of simple sugars from starch	Amylases from fungi and plants.	Production of sugars from starch, such as in making high-fructose corn syrup. In baking, catalyze breakdown of starch in the flour to sugar. Yeast fermentation of sugar produces the carbon dioxide that raises the dough.
	Proteases	Biscuit manufacturers use them to lower the protein level of flour.
Baby foods	Trypsin	To predigest baby foods.
Brewing industry 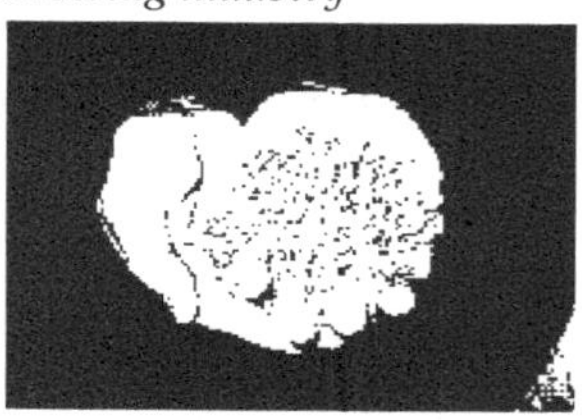 Germinating barley used for malt.	Enzymes from barley are released during the mashing stage of beer production.	They degrade starch and proteins to produce simple sugar, amino acids and peptides that are used by yeast for fermentation.
	Industrially produced barley enzymes	Widely used in the brewing process to substitute for the natural enzymes found in barley.
	Amylase, glucanases, proteases	Split polysaccharides and proteins in the malt.
	Betaglucanases and arabinoxylanases	Improve the wort and beer filtration characteristics.
	Amyloglucosidase and pullulanases	Low-calorie beer and adjustment of fermentability.
	Proteases	Remove cloudiness produced during storage of beers.
	Acetolactatedecarboxylase (ALDC)	Increases fermentation efficiency by reducing diacetyl formation.
Fruit juices	Cellulases, pectinases	Clarify fruit juices
Dairy industry  Roquefort cheese	Rennin, derived from the sto-machs of young ruminant ani-mals (like calves and lambs).	Manufacture of cheese, used to hydrolyze protein.
	Microbially produced enzyme	Now finding increasing use in the dairy industry.
	Lipases	Is implemented during the production of Roquefort cheese to enhance the ripening of the blue-mould cheese.
	Lactases	Break down lactose to glucose and galactose.

Meat tenderizers	Papain	To soften meat for cooking.
Starch industry 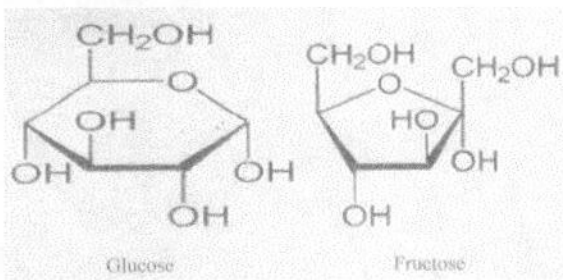 Glucose Fructose	Amylases, amyloglucosid-eases and glucoamylases	Converts starch into glucose and various syrups.
	Glucose isomerase	Converts glucose into fructose in production of high fructose syrups from starchy materials. These syrups have enhanced sweetening properties and lower calorific values than sucrose for the same level of sweetness.
Paper industry  A paper mill	Amylases, Xylanases, Cellulases and ligninases	Degrade starch to lower viscosity, aiding sizing and coating paper. Xylanases reduce bleach required for decolorising; cellulases smooth fibers, enhance water drainage, and promote ink removal; lipases reduce pitch and lignin-degrading enzymes remove lignin to soften paper.
Biofuel industry	Cellulases	Used to break down cellulose into sugars that can be fermented .
	Ligninases	Use of lignin waste
Biological detergent	Primarily proteases, produced in an extracellular form from bacteria	Used for presoak conditions and direct liquid applications helping with removal of protein stains from clothes.
	Amylases	Detergents for machine dish washing to remove resistant starch residues.
	Lipases	Used to assist in the removal of fatty and oily stains.
	Cellulases	Used in biological fabric conditioners.
Contact lens cleaners	Proteases	To remove proteins on contact lens to prevent infections.
Rubber industry	Catalase	To generate oxygen from peroxide to convert latex into foam rubber.
Photographic industry	Protease (ficin)	Dissolve gelatin off scrap film, allowing recovery of its silver content.
Molecular biology 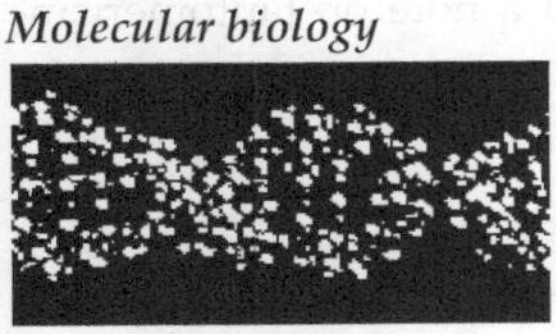 Part of the DNA double helix	Restriction enzymes, DNA ligase and polymerases	Used to manipulate DNA in genetic engineering, important in pharmacology, agriculture and medicine. Essential for restriction digestion and the polymerase chain reaction. Molecular biology is also important

REVIEW QUESTIONS

1. What are carbohydrates? What functions do they serve in the living system?
2. Give the classification of carbohydrates.
3. Write short notes on:

(*a*) Monosaccharides (*b*) Disaccharides
(*c*) polysaccharides

4. Give the structural formula of following:
   (*a*) chitin (*b*) maltose
   (*c*) lactose
5. Write illustrative note on optical isomerism. Explain with example.
6. Write short notes on:
   (*a*) Mutarotation (*b*) Glycosidic bond
   (*c*) Sugar derivatives
7. What are lipids? How are they classified?
8. Discuss the characteristics and functions of a class of lipids having nitrogen base and phosphoric acid as constituents.
9. With structures give examples of two unsaturated fatty acids essential for our body.
10. Describe the structure of steroids. Add a note on the functions of cholesterol.
11. Describe the structure and functions of phospholipids.
12. Write an account on derived lipids.
13. Discus the saturated fatty acids of biological importance.
14. Write short notes on:
    (*a*) Cholesterol (*b*) Phospholipids
    (*c*) Lipoproteins (*d*) Glycolipids
15. Write short notes on:
    (*a*) Zwitterion (*b*) Isoelectric pH
16. What are proteins? Describe briefly the biological importance of proteins.
17. What is the primary structure of proteins? Discuss the biochemical functions of proteins.
18. Discuss in brief the classification and biological importance of proteins.
19. Discuss the secondary structure of proteins.
20. What are the essential and nonessential amino acids?
21. Write short note on quaternary structure of protein.
22. Describe the Watson and Crick model of DNA.
23. Write an account of structure, function and nomenclature of nucleotides.
24. Describe the structure of nitrogenous bases present in nucleic acid. Add a note on tautomerism.
25. Name different types of RNA and discuss their functions.
26. Write short notes on:
    (*a*) Different forms of DNA (*b*) Purine bases of plants
    (*c*) Complimentary base pairs (*d*) Hydrogen bonds in DNA
    (*e*) Ribose and deoxyribose (*f*) 3′–5′ phosphodiester bridge
27. What are enzymes? Describe their classification.
28. Describe the nomenclature of enzymes.
29. Describe the Fischer lock-and-key hypothesis for enzyme specificity.
30. Describe the Koshland induced-fit hypothesis for enzyme specificity.
31. Discuss in detail the industrial applications of enzymes.
32. Give the enzyme purification processes in detail.

33. What are coenzymes? Write briefly on the role of coenzyme in enzyme action.
34. Discuss the various chromatographic techniques for enzyme purification.
35. Define enzyme. What is mechanism of their action? Discuss the factors which influence enzyme action.
36. Give the properties of enzymes.
37. Write short notes on:
    (*a*) Coenzyme
    (*b*) Cofactor
    (*c*) holoenzyme

3

THE CELL

3.1 INTRODUCTION

The **cell** (from Latin *cellulae*: "little rooms") is the structural and functional units of all living organisms. The living matter is composed of mainly six elements like; carbon, hydrogen, oxygen, nitrogen, phosphorus and sulpher. These elements mutually constitute about 90% of the dry weight of the human body. Several other important elements are also found in the cells. These includes Ca, K, Fe, I, F, Zn, Na, Cl, Mg, Cu, Mo, Co and Se.

Carbon is the most leading and versatile element of life. It possesses a unique property to form infinite number of compounds. This is attributed to the ability of carbon to form stable covalent bonds and C-C chains of unlimited length. It is estimated that about 90% of compounds found in living system contain carbon. All living cells that are capable of reproducing themselves have certain basic features in common:

- A membrane, which envelopes the cell, separates its interior from the surroundings, strictly controls what moves in and out and maintains the electric potential of the cell.
- A salty cytoplasm (the substance which makes up most of the cell volume).
- DNA, the hereditary material of genes, which direct the operations of the cell.
- RNA, through which DNA instructions are expressed.
- Enzymes and other protein machinery.
- A variety of biomolecules.
- The capacity to divide by mitosis.
- Metabolism, including the taking in of raw material, using it to build cell components, or breaking it down for energy, and releasing by-products.
- Protein biosynthesis.
- The ability to respond to external and internal stimuli.

These functions and abilities are expressed in the **cell cycle** : the "birth", growth, reproduction, and "death" of individual cells.

Chemical Organization of Cells

The chemical organization of cells consists of chemical components. The chemical components of cell can be classified in two types:

- Inorganic component (water and mineral salts)
- Organic components (carbohydrates, protein, lipids, nucleic acids) (as discussed in chapter no. 2)

Water and Salts

Water serves as a natural solvent for mineral ions and other substances and also as a dispersion medium of the colloid system of protoplasm. Water molecules also participate in many enzymatic reactions in the cell. Water exists in the cell in two forms, *i.e.*, free and bound. Free water represents 95% of the total cellular water and it is the principal part used as a solvent for solutes. Bound water, which represents 4–5% of the total cellular water, it is loosely held to the proteins by hydrogen bonds and other forces. Water is also used to eliminate substances from the cell and to absorb heat; thus preventing drastic temperature changes in the cell. Another important property of water is its ionization in to a hydroxide anion and a hydrogen ion.

Salts dissociated in to anions and cations are important in maintain osmotic pressure and the acid base equilibrium of the cell. Some of the inorganic ions, such as magnesium, are indispensable as cofactor in enzymatic activities; others, such as inorganic phosphate, participate to form ATP. Certain mineral ions are also found in cells as a part of larger macromolecule. The importance of some inorganic ions in enzymes is given in following Table 3.1:

Table 3.1

Ions	Functions in cell
Calcium	Calmodulin, actomyosin, ATPase
Cobalt	Peptidases
Zinc	Carbonic anhydrase, peptidase, transcription factor III A, ADH
Copper	Tyrosinase, ascorbic acid oxidase
Manganese	Peptidases
Magnesium	Chlorophyll, phosphatses
Iron	Haemoglobin, cytochromes, peroxidases
Molybdenum	Nitrate reductase, xanthine oxidase

Cell Theory

Wilson and Morrison (1966) tried to suggest that the cell is an integrated and continuously changing system and, when it ceases to be a cell in biological sense, it becomes merely a mass disintegrated matter. In order that we understand the present status and the problems of the cells, it is desirable to know the history. The seeds of progress have frequently been sown in the past, often unconscientiously, and history should be read not only for pleasure but for profit also. In 1665, Robert Hooke introduced the term *cell* while describing microscopic texture of the cork (Fig. 3.1). This was followed by illustrations conforming to cellular nature of living organisms by a number of workers during the next two centuries. However, Schieiden and Schwann are generally accredited as the exponents of the cell theory in 1838, according to which both plants and animals are made up of aggregations of basic units called cells, each consisting of a nucleus (kern) and cytoblastema (cytoplasm). The theory given by Schieiden and Schwann in the nineteenth century introduced the cell doctrine, which may be summarized as:

- Cells of organisms are differentiated into distinct cell type.
- A nucleus is a feature of all cells with the exception of viruses and red blood cells.
- The cytoplasm is the living content of the cell and determines its activities, and thus the whole organism.
- New cells originate from pre-existing cells through division. In any case, cells never arise *de novo*.

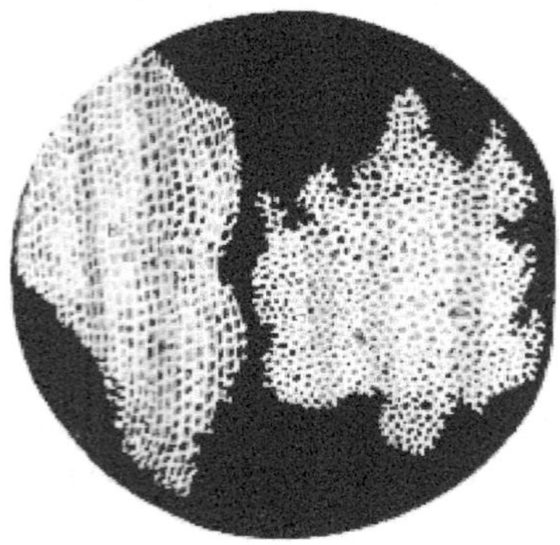

Fig. 3.1 The structure of cork

The cells of the living kingdom may be divided in to two categories on the bases of their structural organization *i.e.*, ***prokaryotic*** and ***eukaryotic***. The *prokaryotes* include all bacteria and blue green algae, while *eukaryotes* consist of protists, animals and plants. In addition to these two groups (*prokaryotes and eukaryotes*), we can consider viruses (discuss later half) too as quasi-living units. In reality, a virus is neither an organism nor a cell; it consists of only a core of nucleic acid an extra coat of protein.

3.1.1 Prokaryotic Cells

The word *prokaryote* means 'primitive nucleus' in Greek, and has the following features:

- The cell wall of prokaryotes consists of ***peptidoglycan*** in bacteria, and acts as an additional barrier against exterior forces. It also prevents the cell from "exploding" from osmotic pressure against a hypotonic environment.
- The *cytoplasm* of prokaryotes (the liquid which makes up most of the cell volume) is diffuse and granular due to *ribosomes* (protein factories) floating in the cell.
- The *plasma membrane* (a phospholipid bilayer) separates the interior of the cell from its environment and serves as a filter and communications beacon.
- A prokaryotic chromosome is usually a circular molecule (an exception is that of the bacterium *Borrelia burgdorferi*, which causes Lyme disease). Even without a real *nucleus*, the DNA is somehow condensed in a *nucleoid*.
- Prokaryotes can carry extra chromosomal DNA elements called *plasmids*, which are usually circular. Plasmids can carry additional functions, such as antibiotic resistance.
- Some prokaryotes have *flagella* which facilitate them to move actively instead of passively drifting.

- Most prokaryotes range from 0.5–3 μm in diameter and 2–8 μm in length. They have a volume of 10^{-12} ml / cell and may weigh approximately 10^{-12} g/ cell.
- They are variously shaped like cocci (meaning berry), spherical to oval, or may be found as tetrads or as a bunch of grapes, or may be rod shaped bacillus or may be spiral.

External Attachments

These attachments are present external to the cell wall, and include glycocalyx, flagella, axial filaments, fimbriae and pili.

Cell Wall

The bacterial cell wall is complex, semi-rigid and is responsible for characteristic shape. It protects cell organelles from external environment. The cell wall is composed of a macromolecular network called peptidoglycan which consists of a repeated disaccharide (N-acetyl glucosamine (NAG) and N-acetyl muramic acid (NAM) attached by polypeptides to form a lattice (Fig. 2.17). Alternating NAM and NAG molecules are linked in rows of 10–65 sugars to form a carbohydrate backbone (glycan portion of peptidoglycan). The adjacent rows are linked by polypeptides (peptide portion of peptidoglycan).

Certain primitive group of prokaryotes does not possess at all are possesses a very little cell wall material, For example. Mycoplasma, which are the smallest bacteria. They posses unique plasma membrane having lipids known as sterol, which help in prevention of osmotic lysis. Archaebacteria lack or have unusual cell walls of polysaccharides and proteins without peptidoglycan or contain psedomurein. Psedomurein is made up of N-acetyl alosaminuronic acid instead of NAM and lacks D-amino acids. Gram positive bacterial cell wall consists of thick, rigid structure due to several peptidoglycan layers, and contains teichoic acids. Gram positive bacterial cell wall contains a few layers of peptidoglycan and outer membrane which is made up of lipoprotein, liposachharides and phospholipids.

Cytoplasmic Membrane

It is a thin phospholipid bilayer surrounding the cell (about 8 nm thick) acting as a barrier from external environment. It is a highly selective membrane enabling the cell to concentrates specific metabolites and excretes waste. It is primarily consists of phospholipids and proteins, and lacks sterols, and hence it is less rigid than eukaryotic organisms. The membrane has phospholipid bilayer and proteins embedded in it with hydrophobic region towards inside. Hydrophilic region is exposed to the external environment. The overall structure of the cytoplasmic membrane is stabilize by hydrogen bonds and hydrophobic interactions, cation like Ca^{2+} and Mg^{2+} ionic bonds with negatively charged phospholipids. Although, the cytoplasmic membrane appears rigid, it is actually fluid in nature, and allows phospholipid and protein molecules to have freedom to move about the membrane. The membrane has selective permeability and prevents destruction by anti-microbial agents. It permits active and passive transport of nutrients.

Cytoplasm

It is the polyphasic, colloidal, living system of cell, which is thick, semitransparent elastic in nature, and has higher water content. It contains hydrophilic components (protein particles, carbohydrates and salts) and hydrophobic components (globules of lipid or fats).

- Sol-gel phase reversal during metabolic activities
- Molecules form large aggregates due to cohesive forces.
- Absorption and excretion.
- Cytoplasm shows resistance towards flow, which makes the particles to be in a state of suspension rather than settling down due to gravitational pull
- It exhibits Brownian movement.

Nuclear Area

Prokaryotes do not have nuclear envelope and do not show cytoplasmic organelles. The bacterial shows single chromosome in cytoplasm which consists of single, circular super coiled naked DNA. This closed loop of DNA contains hereditary information of cell, and this region is referred to as nucleoid or genophore, which is exclusively made of DNA. There is total absence of nuclear membrane in prokaryotes, and the available evidence shows that the chromatin body is equivalent to a single chromosome, hence termed as genophore. About 1/5 of the cell interior of *E. Coli.* Is occupied by DNA and it contains genetic information to code for about 3000 different proteins. Some bacteria have mesosomes, to which DNA may be attached, and which also appear to carry respiratory components. Smaller molecules of DNA exist apart from chromosome loops called as plasmids, contain few genes which are not essential for growth and metabolism, but confer special adaptive methods for better survival. Ribosomes are usually present in cytoplasm and are more active in protein-synthesizing cells. They contain RNA and protein, and are essential machinery for protein synthesis. Prokaryotic ribosome (70s) comprises two subunits-50s and 30s. 50s subunit is spherical to dome-shaped in structure, and 30s subunit is ellipsoid to prolate in structure.

Inclusion Bodies

The prokaryotic cytoplasm has many inclusions, commonly called as inclusion bodies which store nutrients for starvation periods. Some of these are as follows:

- Polysaccharide granules
- Sulphur granules
- Gas vacuoles
- Carboxysomes
- Magnetosomes
- Metachromatic granules
- Lipid inclusions/granules

Endospores

Mostly Gram +ve bacteria produce highly resistant special structures called Endospores.

These organisms grow, mature and reproduce for several hours as vegetative cells. Endospores contain a little amount of water and exhibit very few reactions. They have a large amount of dipicolinic acid which stabilizes the proteins. The spore formation begins under unfavourable conditions. The bacterial chromosomes replicates, a small amount of cytoplasm gathers with it, and the membrane grows to seal of the developing spore; then thick layers of peptidoglycan coat it and vegetative cell degenerate. On return on favourable conditions, protective layers breakdown, and spores germinate to vegetative cells. Thick layers of peptidoglycan form and coat it, and the vegetative cells disintegrate. As a result, the spore is freed.

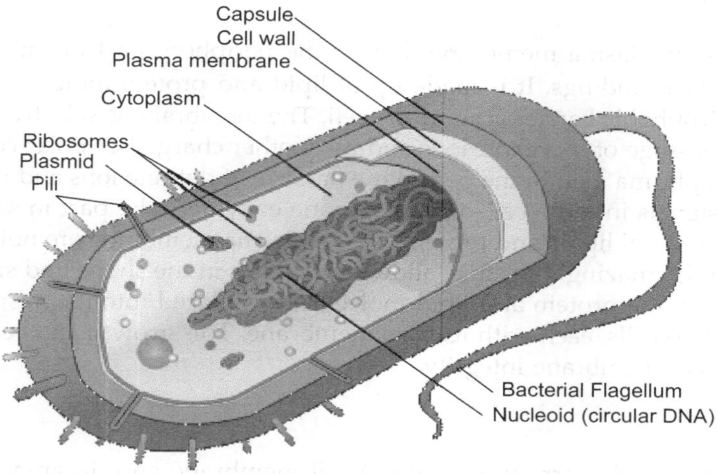

Fig. 3.2 A typical prokaryotic cell

3.1.2 Eukaryotic Cells

The word eukaryotic means 'true nucleus'. Eukaryotic cells are larger and more complex than the prokaryotic cell. They have organelles with their own cell membranes. Single-celled eukaryotic organisms are very diverse, but many colonial and multicellular forms also exist. (The multicellular kingdoms: Animalia, Plantae and Fungi, are all eukaryotic). Eukaryotes consist of three main components:

1. Cell membrane
2. Cytoplasm
3. Nucleus

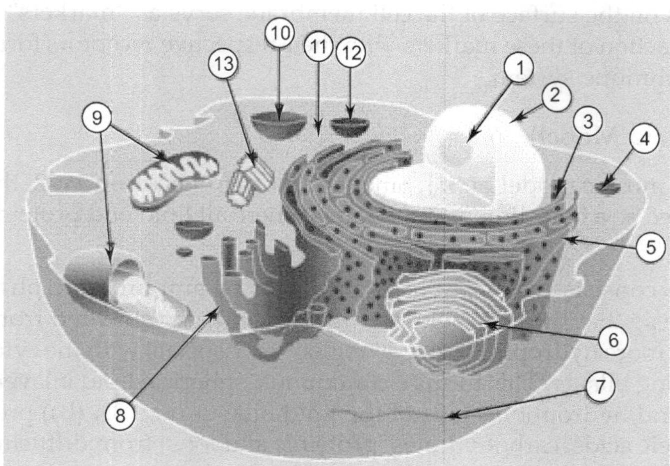

Fig. 3.3 Schematic of typical animal cell, showing subcellular components. (1) Nucleolus (2) Nucleus (3) Ribosome (4) Vesicle (5) Rough endoplasmic reticulum (ER) (6) Golgi apparatus (7) Cytoskeleton (8) Smooth ER (9) Mitochondria (10) Vacuole (11) Cytoplasm (12) Lysosome (13) Centrioles

1. Cell Membrane

The cell membrane or plasma membrane defines the periphery part of the cell, separating its contents from the surroundings. It is made up of lipid and protein molecules that form a thin, tough, pliable, hydrophobic barrier around the cell. The membrane is selectively permeable, and barrier to the free passage of inorganic ions and most other charged or polar compounds. Transport proteins in the plasma membrane permit the passage of definite ions and molecules; receptor proteins transmit signals into the cell; and membrane enzymes take part in some reaction pathways. Since the individual lipids and proteins of the plasma membrane are not covalently linked, the whole structure is amazingly flexible, allowing changes in the shape and size of the cell. As a cell grows, recently made protein and lipid molecules are placed into its plasma membrane; cell division produces two cells, each with its own membrane. This growth and cell division (fission) occurs without loss of membrane integrity.

Functions

The protoplasm of a cell is surrounded by the cell membrane and, in animal cells, physically separates the intracellular components. In fungi, some bacteria, and plants, an additional cell wall forms the outermost boundary; however, the cell wall plays mostly a mechanical support role rather than a role as a selective boundary. The cell membrane also plays a role in anchoring the cytoskeleton to provide shape to the cell, and in attaching to the extracellular matrix and other cells to help group cells together to form tissues. The barrier is differentially permeable and able to regulate what enters and exits the cell, thus facilitating the transport of materials needed for survival. The movement of substances across the membrane can be either passive, occurring without the input of cellular energy, or active, requiring the cell to expend energy in moving it. The membrane also maintains the cell potential. Specific proteins embedded in the cell membrane can act as molecular signals that allow cells to communicate with each other. Protein receptors are found ubiquitously and function to receive signals from both the environment and other cells. These signals are transduced and passed in a different form into the cell. For example, a hormone binding to a receptor could open an ion channel in the receptor and allow calcium ions to flow into the cell. Other proteins on the surface of the cell membrane serve as "markers" that identify a cell to other cells. The interaction of these markers with their respective receptors forms the basis of cell-cell interaction in the immune system.

Structure (Fluid Mosaic Model)

According to the fluid mosaic model of S. J. Singer and Garth Nicolson 1972, the biological membranes can be considered as a two-dimensional liquid where all lipid and protein molecules diffuse more or less easily.

The cell membrane consists primarily of a thin layer of amphipathic phospholipids which spontaneously arrange so that the hydrophobic "tail" regions are shielded from the surrounding polar fluid, causing the more hydrophilic "head" regions to associate with the cytosolic and extracellular faces of the resulting bilayer. This forms a continuous, spherical lipid bilayer. The arrangement of hydrophilic heads and hydrophobic tails of the lipid bilayer (Fig. 3.4 (B)) prevent polar solutes (*e.g.*, amino acids, nucleic acids, carbohydrates, proteins, and ions) from diffusing across the membrane, but generally allows for the passive diffusion of hydrophobic molecules. This affords the cell the ability to control the movement of these substances via transmembrane protein complexes such as pores and gates. Membranes serve diverse functions in eukaryotic and prokaryotic cells. One important role is to regulate the movement of materials into and out of cells. The phospholipid bilayer structure (fluid mosaic model) with specific membrane proteins accounts for the selective permeability of the membrane and passive and active transport mechanisms. In addition, membranes in prokaryotes and in the mitochondria and chloroplasts of eukaryotes facilitate the synthesis of ATP through chemiosmosis. The cell membrane contains many integral membrane proteins, which infuse the entire surface. The structures can be visualized by electron microscopy or fluorescence microscopy. It can be found on the inside of the membrane, the outside, or membrane spanning. These may include integrins, cadherins, desmosomes, clathrin-coated pits, caveolaes, and different structures involved in cell adhesion.

Membrane Skeleton

The cytoskeleton is found underlying the cell membrane in the cytoplasm and provides a scaffolding for membrane proteins to anchor to, as well as forming organelles that extend from the cell. Indeed, cytoskeletal elements interact extensively and intimately with the cell membrane. Anchoring proteins restricts them to a particular cell surface, for example, the apical surface of epithelial cells that line the vertebrate gut, and limits how far they may diffuse within the bilayer. The cytoskeleton is able to form appendage-like organelles, such as cilia, which are microtubule-based extensions covered by the cell membrane, and filopodia, which are actin-based extensions. These extensions are ensheathed in membrane and project from the surface of the cell in order to sense the external environment and/or make contact with the substrate or other cells. The apical surfaces of epithelial cells are dense with actin-based finger-like projections known as microvilli, which increase cell surface area and thereby increase the absorption rate of nutrients.

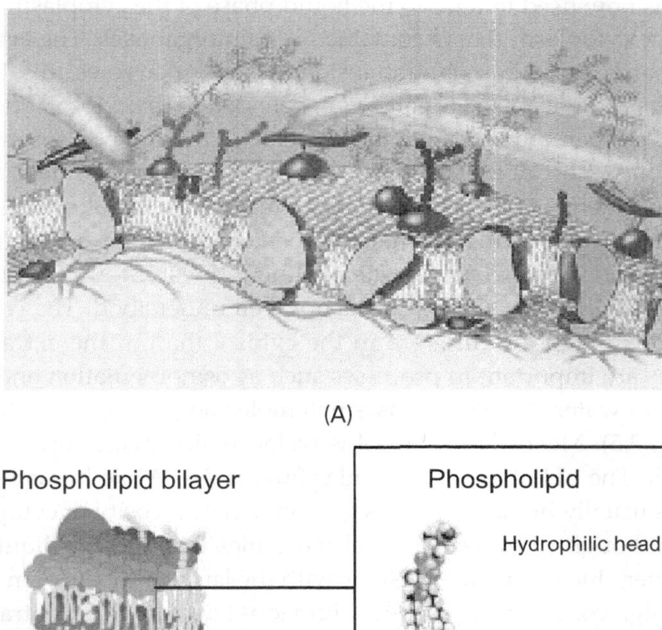

(A)

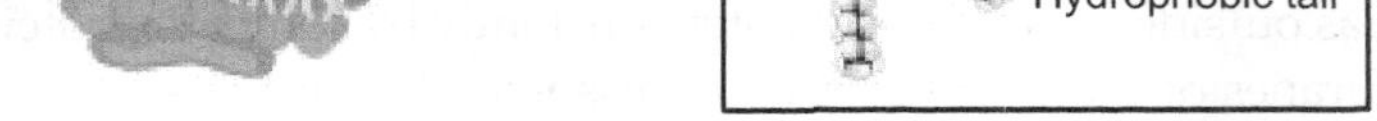

(B)

Fig. 3.4 (*a*) "Fluid mosaic" model of the structure of plasma membrane showing the organiza-tion of lipid bilayers and embedded proteins and carbohydrates. (*b*) Phospholipid showing hydrophilic head and hydrophobic tail

2. Cytoplasm

The cytoplasm is the internal volume of a cell that is bounded by the plasma membrane. In eukaryotic cells, cytoplasm is the viscid, semi fluid matter contained within the plasma membrane of a biological cell; the contents of the cell nucleus are not the part of the cytoplasm and are instead called the nucleoplasm. The cytoplasm contains organelles, such as mitochondria, which are filled with liquid that is kept separate from the rest of the cytoplasm by biological membranes. The cytoplasm is the site where most cellular activities occur, such as many metabolic pathways like glycolysis, and processes such as cell division. The inner, granular mass is called the endoplasm and the outer, clear and glassy layer is called the cell cortex or the ectoplasm. The insoluble constituents of the cytoplasm include the organelles and the cytoskeleton, which gives to the cytoplasm a gel-like structure and consistency.

The cytoplasm has major components given below:

- The cytosol
- Organelles

Cytosol

The term cytosol was first introduced in 1965 by H.A. Lardy, and initially referred to the liquid that was produced by breaking cells apart and pelleting all the insoluble components by ultracentrifugation. The term cytosol is now used to refer to the liquid phase of the cytoplasm in an intact cell; this excludes any part of the cytoplasm that is contained within organelles. The cytosol or intracellular fluid (or cytoplasmic matrix) is the liquid found inside cells. In eukaryotes, this liquid is separated by cell membranes from the contents of the organelles suspended in the cytosol. The entire contents of a eukaryotic cell within cell membrane, minus the contents of the cell nucleus, are referred to as the cytoplasm. In prokaryotes, most of the chemical reactions of metabolism take place in the cytosol, while a few take place in membranes or in the periplasmic space. In eukaryotes, while many metabolic pathways still occur in the cytosol, others are contained within organelles. The cytosol is a complex mixture of substances dissolved in water. Although water forms the large majority of the cytosol, its structure and properties within cells is not well understood. The concentrations of ions such as sodium and potassium are different in the cytosol than in the extracellular fluid; these differences in ion levels are important in processes such as osmoregulation and cell signalling. The cytosol consists mostly of water, dissolved ions, small molecules, and large water-soluble molecules (such as proteins) (Fig. 3.5). Most of the cytosol is water, which makes up about 70% of the total volume of a typical cell. The pH of the extracellular fluid is 7.4. While human cytosolic pH ranges between 7.0–7.4, and is usually higher if a cell is growing. The viscosity of cytoplasm is roughly the same as pure water, although diffusion of small molecules through this liquid is about fourfold slower than in pure water, due mostly to collisions with the large numbers of macromolecules in the cytosol. Cytosol has a high concentration of potassium ions and a low concentration of sodium ions. This cytosol difference in ion concentrations is critical for osmoregulation, since if the ion levels were the same inside a cell as outside, water would enter constantly by osmosis. Protein molecules that do not bind to cell membranes or the cytoskeleton are dissolved in the cytosol.

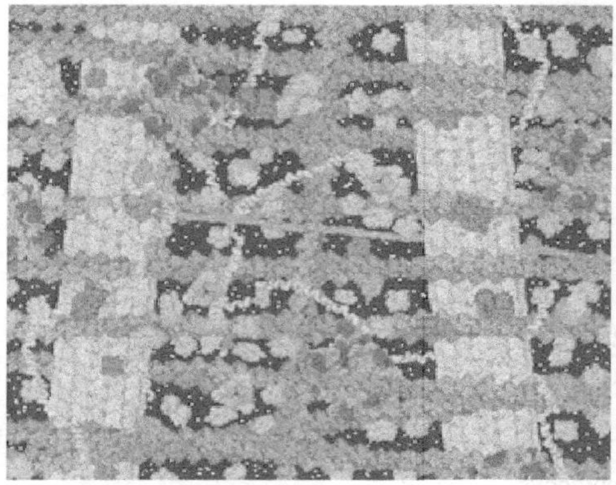

Fig. 3.5 The cytosol is a crowded solution of many different types of molecules that fills much of the volume of cells

Function of Cytosol

The cytosol has performed many functions and is instead the site of multiple cell processes. Examples of these processes include signal transduction from the cell membrane to sites within the cell, such as the cell nucleus, or organelles. Another major function of cytosol is to transport metabolites from their site of production to where they are used. This is relatively simple for water-soluble molecules, such as amino acids, which can diffuse rapidly through the cytosol. However, hydrophobic molecules, such as fatty acids or sterols, can be transported through the cytosol by specific binding proteins, which shuttle these molecules between cell membranes. Molecules taken into the cell by endocytosis or on their way to be secreted can also be transported through the cytosol inside vesicles, which are small spheres of lipids that are moved along the cytoskeleton by motor proteins. The cytosol is the site of most metabolisms in prokaryotes, and a large proportion of the metabolism of eukaryotes. For instance, in mammals about half of the proteins in the cell are localized to the cytosol. The most complete data are available in yeast, where metabolic reconstructions indicate that the majority of both metabolic processes and metabolites occur in the cytosol. Major metabolic pathways that occur in the cytosol in animals are protein biosynthesis, the pentose phosphate pathway, glycolysis and gluconeogenesis.

Organelles

Organelles are membrane-bound compartments within the cell that have specific functions. Some major organelles that are suspended in the cytosol are the mitochondria, the endoplasmic reticulum, the Golgi apparatus, ribosomes, vacuoles, lysosomes, centriole and in plant cells chloroplasts.

Mitochondria

The word mitochondrion comes from the Greek *mitos*, thread or *chondrion*, granule. Mitochondria are found in both animals and plants. The size of mitochondria depends upon the functional stage of cell. The width is about stable (0.5μ–1.0μ) but the length varies significantly measuring from 1.5–10μ. It is powerhouses of the eukaryotic cell. Mitochondria pump out energy by converting glucose to ATP (adenosine triphosphate). ATP acts like a battery of sorts, storing energy until it's needed for day to-day living. In addition to supplying cellular energy, mitochondria are involved in a range of other processes, such as cellular differentiation, cell death, signalling, as well as the control of the cell cycle and cell growth. Mitochondria have been implicated in several human diseases, including mitochondrial disorders and cardiac dysfunction, and may play a role in the aging process. The structure of mitochondria includes outer membrane, the intermembrane space, the inner membrane, and the cristae and matrix (Fig. 3.6).

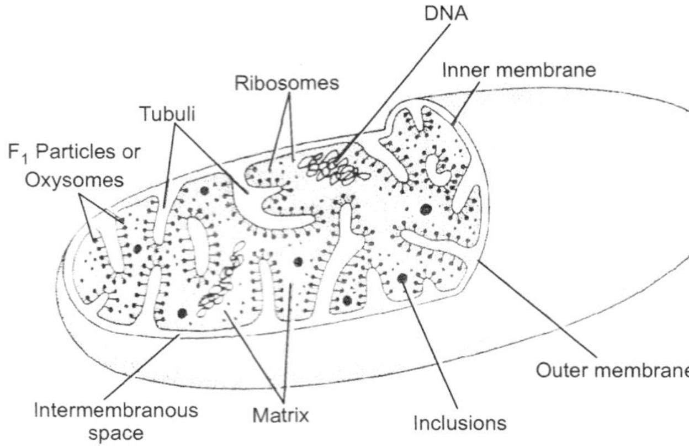

Fig. 3.6 The structure of mitochondrion

Outer Membrane

The total organelle enclosed by the outer mitochondrial membrane. It has a protein-to-phospholipid ratio analogous to that of the eukaryotic plasma membrane. The mitochondrial outer membrane can link with the ER membrane, in a structure called MAM (mitochondria-associated ER-membrane). This is significant in ER-mitochondria calcium signalling and involved in the transfer of lipids among the ER and mitochondria.

The Intermembrane Space

The space amid the outer membrane and the inner membrane is called intermembrane space. As the outer membrane is freely permeable to small molecules, the concentrations of small molecules such as ions and sugars in the intermembrane space are the same as the cytosol.

The Inner Mitochondrial Membrane

The finger like projection of mitochondrial inner membrane is called cristae, which elongate in to the matrix. The outer face of inner membrane directed towards the outer chamber is known as cytosol face (C-face), and the inner face of inner membrane directed towards matrix is known as matrix face (M-face). The inner mitochondrial membrane. contains more than 150 different polypeptides, and has a very high protein-to-phospholipid ratio (more than 3:1 by weight, which is about 1 protein for 15 phospholipids). The inner mitochondrial membrane having around 1/5 part of the total protein. In addition, the inner membrane is rich in an unusual phospholipid, cardiolipin. This phospholipid was originally discovered in beef hearts in 1942, and is usually characteristic of mitochondrial and bacterial plasma membranes. Cardiolipin contains four fatty acids rather than two and may help to make the inner membrane impermeable. The inner mitochondrial membrane contains proteins with five types of functions:

- It carried out the redox reactions of oxidative phosphorylation.
- Inner mitochondrial membrane is rich in enzyme ATPase synthase, which generates ATP in the matrix.
- Metabolite passage into and out of the matrix is regulated by specific transport proteins, Permease.
- It acts as protein import machinery.

Cristae: The inner mitochondrial membrane is thrown into folds or invagination called cristae, which expand into the matrix. The cristae are irregularly shaped; some are villose while others are finger like. They may be parallel to the long axis, perpendicular to the long axis of mitochondria. These irregular forms divide the central space or matrix into interconnected chambers. The outer face of inner membrane directed towards the outer chamber is called C-face (cytosol face). The inner face of inner membrane, directed towards matrix is called M-face (matrix face).

Matrix: The matrix is the space covered by the inner membrane. It contains about 2/3 of the total protein in a mitochondrion. The matrix play significant role in the production of ATP with assist of the ATP synthase contained in the inner membrane. The matrix contains highly concentrated enzymes, few lipids, particular mitochondrial ribosomes, tRNA, and various copies of the mitochondrial DNA genome. The key functions of the enzymes, include oxidation of pyruvate, fatty acids, and citric acid cycle.

Function

The most prominent roles of mitochondria are to produce ATP (*i.e.*, phosphorylation of ADP) through respiration, and to regulate cellular metabolism. The central set of reactions involved in ATP production is collectively known as the citric acid cycle, or the Krebs cycle. However, the mitochondrion has many other functions in addition to the production of ATP.

Endoplasmic Reticulum

Eukaryotic cells have numerous interrelated membrane bound compartments usually known as endoplasmic reticulum (ER) (Fig. 3.7), which plays significant role in cellular activities. There are several components of the endomembrane system, including mitochondria, Golgi, chloroplast, microbodies, and vacuoles etc., embedded in the cytoplasmic matrix. Rough endoplasmic reticulum is well developed in cells that are synthesize proteins, while smooth endoplasmic reticulum synthesize lipids and steroids, metabolize carbohydrates and steroids, and regulate calcium concentration, drug detoxification, and attachment of receptors on cell membrane proteins. The general structure of the endoplasmic reticulum is an extensive membrane network of cisternae (sac-like structures) held together by the cytoskeleton. The phospholipid membrane encloses a space, the cisternal space (or lumen), from the cytosol, which is continuous with the perinuclear space. The functions of the endoplasmic reticulum vary greatly depending on the exact type of endoplasmic reticulum and the type of cell in which it resides. The three varieties are called rough endoplasmic reticulum, smooth endoplasmic reticulum and sarcoplasmic reticulum.

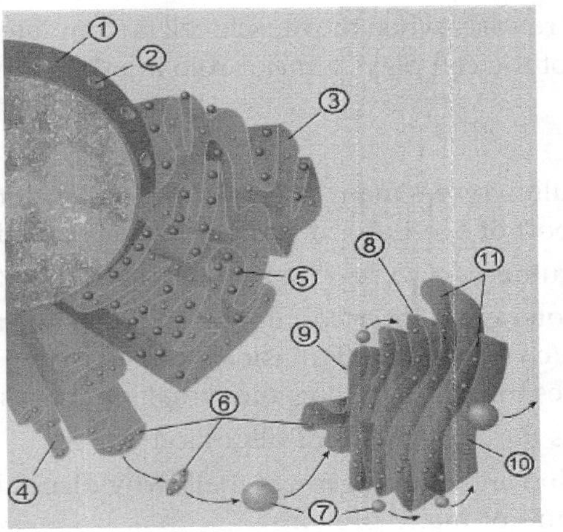

Fig. 3.7 (1) Nucleus (2) Nuclear pore (3) Rough endoplasmic reticulum (RER) (4) Smooth endoplasmic reticulum (SER) (5) Ribosome on the rough (ER) (6) Proteins that are transported (7) Transport vesicle (8) Golgi apparatus (9) Cis face of the Golgi apparatus (10) Trans face of the Golgi apparatus (11) Cisternae of the Golgi apparatus

Rough Endoplasmic Reticulum

The Rough Endoplasmic Reticulum (RER) is well developed and its membranes are studded with ribosomes which are associated with protein synthesis. Though, the ribosomes bound to the RER at any one time are not an established component of this organelle's structure as ribosomes are constantly being bound and released from the membrane. A ribosome only binds to the ER once it begins to synthesize a protein destined for the secretory pathway. Here, a ribosome in the cytosol begins synthesizing a protein until a signal recognition particle recognizes the pre-piece of 5–15 hydrophobic amino acids preceded by a positively charged amino acid. This signal sequence allows the recognition particle to bind to the ribosome, causing the ribosome to bind to the RER and pass the new protein through the ER membrane. The pre-piece is then cleaved off within the lumen of the ER and the ribosome released back into the cytosol.

Smooth Endoplasmic Reticulum

The membranes of Smooth Endoplasmic Reticulum (SER) are having lack ribosomes. The Smooth Endoplasmic Reticulum (SER) has functions in several metabolic processes, including synthesis of lipids and steroids, metabolism of carbohydrates, regulation of calcium concentration, drug detoxification, attachment of receptors on cell membrane proteins, and steroid metabolism. It is connected to the nuclear envelope. Smooth endoplasmic reticulum is present in a diversity of cell type both animal and plant and it serves diverse functions in each. The Smooth ER also contains the enzyme glucose-6-phosphatase which converts glucose-6-phosphate to glucose, a step in gluconeogenesis. The SER consists of tubules and vesicles that branch forming a network. In some cells there are dilated areas like the sacs of RER. The network of SER allows increased surface area for the action or storage of key enzymes and the products of these enzymes.

Sarcoplasmic Reticulum

The Sarcoplasmic Reticulum (SR), from the Greek sarx, ("flesh"), is a special type of smooth ER found in smooth and striated muscle. The only structural difference between this organelle and the SER is the medley of proteins they have, both bound to their membranes and drifting within the confines of their lumens. This fundamental difference is indicative of their functions: the SER synthesizes molecules while the SR stores and pumps calcium ions. The SR contains large stores of calcium, which it sequesters and then releases when the muscle cell is stimulated. The SR's release of calcium upon electrical stimulation of the cell plays a major role in excitation-contraction coupling.

Functions

- The endoplasmic reticulum serves many general functions, including the facilitation of protein folding and the transport of synthesized proteins in sacs called cisternae.
- The endoplasmic reticulum is also part of a protein sorting pathway.
- Integral membrane proteins are inserted into the endoplasmic reticulum membrane as they are being synthesized (co-translational translocation). Insertion into the endoplasmic reticulum membrane requires the correct topogenic signal sequences in the protein.
- Glycosylation involves the attachment of oligosaccharides.
- Disulfide bond formation and rearrangement that's why disulfide bonds stabilize the tertiary and quaternary structure of many proteins.
- Drug Metabolism: The smooth ER is the site at which some drugs are modified by microsomal enzymes which include the cytochrome P450 enzymes.

Golgi Apparatus

The Golgi apparatus (also Golgi body, Golgi complex) is derived from the endomembrane system found in most eukaryotic cells. It was recognized in 1898 by the Italian physician Camillo Golgi and was named after him. The most important function of the Golgi apparatus is to process and package macromolecules, such as proteins and lipids, after their synthesis. Before synthesis they make their way to their destination; it is mostly important in the processing of proteins for secretion. The Golgi is composed of membrane-bound stacks known as cisternae (singular: cisterna). Between four and eight cisternae are usually present; however, in some protists as many as sixty have been observed. Each cisterna comprises a flattened membrane disk, and carries golgi enzymes to help or to modify proteins that travel through them. They are found in both plant and animal cells. The cisternae stack has four functional regions: the cis-Golgi network, medial-Golgi, endo-Golgi, and trans-Golgi network as shown in above Fig. 3.7. The important functions performed by golgi apparatus are given below:

- The Golgi apparatus is integral in modifying, sorting, and packaging these macromolecules for cell secretion (exocytosis) or use within the cell.
- It primarily modifies proteins delivered from the rough endoplasmic reticulum but is also involved in the transport of lipids around the cell, and the creation of Lysosome.
- The Golgi plays an important role in the synthesis of proteoglycans, which are molecules present in the extracellular matrix of animals.
- It is also a major site of carbohydrate synthesis.
- The Golgi has a putative role in apoptosis.

Lysosomes

Lysosome was discovered by the Belgian cytologist Christian de Duve in 1949. The name lysosome derives from the Greek words lysis, which means dissolution or destruction, and soma, which means body. Lysosomes are spherical organelles that contain hydrolytic enzymes. They are found in animal cells, while lytic vacuoles performed similar roles in yeast and plants cell. They crumble food material; consequently it is easier to digest. Lysosomes digest surplus organelles, food particles, and engulfed viruses or bacteria due to its polymorphic character. The membrane around a lysosome allows the digestive enzymes which work at the acidic pH (4.5). The size of lysosomes varies from 0.1-5.0 µm. Four types of lysosomes have been documented.

(*i*) Primary lysosomes
(*ii*) Secondary lysosomes
(*iii*) Autolysosomes
(*iv*) Telolysosomes

The inner surface of the lysosomes is acidic (pH 4.8) compared to the somewhat alkaline cytosol (pH 7.2). The lysosome maintains this pH differential by pumping protons (H^+ ions) from the cytosol across the membrane via proton pumps and chloride ion channels. Different types of enzymes are present in lysosomes, which facilitate them to digest almost all major chemical constitu ents of the living cell. Some enzymes are listed below:

- Phosphodiesterase, Deoxyribonuclease II, Ribonuclease II, Cholesterol esterase: hydrolases acting on ester bond.

- Nucleases: which digest nucleic acids and hydrolyse them into nitrogenous base, sugar and phosphate.
- Proteases: hydrolyse proteins into amino acids, *e.g.*, peptidase, collagenase and cathepsin enzymes.
- Lipase: This includes acid lipase and phospholipase, *e.g.*, phospholipase, esterase.
- Sulphatases: These hydrolyse sulphate ester into fragments, *e.g.*, arylsulphatase.

The main functions performed by lysosomes are discuss below:

Lysosomes having the capacity to dispose of all waste material from the cell. They are used for the digestion of macromolecules from phagocytosis (ingestion of other dying cells or larger extracellular material, like foreign invading microbes), endocytosis (where receptor proteins are recycled from the cell surface), and autophagy (wherein old or unneeded organelles or proteins, or microbes that have invaded the cytoplasm are delivered to the lysosome). Autophagy may also lead to autophagic cell death, a form of programmed self-destruction, or autolysis, of the cell, which means that the cell is digesting itself. Other functions include digesting foreign bacteria that invade a cell and helping repair damage to the plasma membrane by serving as a membrane patch, sealing the wound. Abnormal behaviour of lysosomes may sometimes cause inborn disorders.

Vacuole

A vacuole is present in all plants and fungal cells and some protists, and bacterial cells as membrane bound organelle. Their occurrence has been reported in animal cells also. Vacuoles are essentially enclosed compartments which are filled with water containing inorganic and organic molecules including enzymes in solution, though in certain cases they may contain solids which have been engulfed. Vacuoles are formed by the fusion of multiple membrane vesicles and are effectively just larger forms of these. The organelle has no basic shape or size, its structure varies according to the needs of the cell. A vacuole is surrounded by a membrane called the tonoplast. Also called the vacuolar membrane, the tonoplast is the cytoplasmic membrane surrounding a vacuole, separating the vacuolar contents from the cell's cytoplasm. As a membrane, it is mainly involved in regulating the movements of ions around the cell, and isolating materials that might be harmful or a threat to the cell. The structure of vacuole is given below:

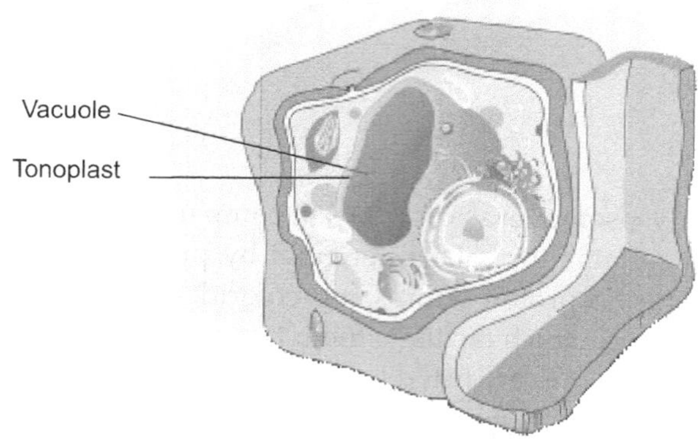

Fig. 3.8 Structure of vacuole in plant cell

Functions of Vacuole

The function and importance of vacuoles varies considerably according to the nature of cell in which they are present. Vacuoles are filled with liquid which is rich in non-colloidal and inorganic substances. The vacuolar fluids have a higher ionic concentration than the cytoplasm. Vacuoles also serve as storage organelles for metabolites or toxic excretory products. Vacuoles also take part in a autophagy, maintaining a equilibrium between biogenesis (production) and degradation (or turnover), of many substances and cell structures in certain organisms. They also assist in damage of invading bacteria or of misfolded proteins that have begun to build up within the cell. In protists, vacuoles have the additional function of storing food which has been absorbed by the organism, and assist in the digestive and waste management process for the cell. In animal cells, vacuoles perform mostly subordinate roles, assisting in larger processes of exocytosis and endocytosis.

Ribosomes

Ribosomes are small nonmembranous and dense spheroidal particles or granules first observed by electron microscopy by Palade in 1955. Ribosomes were so named because of their high contents of ribonucleic acid. These are found in both prokaryotic and eukaryotic cells. These are small globular particles of approximately 250Å diameter. Most of Ribosomes are linked with the outer surface of endoplasmic reticulum, many are free in the cytosol, and some are found in the nucleus and mitochondria (also found in chloroplast in plant cell). On the basis of sedimentation coefficient they are of two types.

(*a*) **70S Ribosomes:** They are smaller in size and their sedimentation coefficient is 70S (found in prokaryotes and chloroplast). Each ribosome is made up of 2 sub unit, one about half the size of the other. In 70S ribosome, the larger subunit is 50S and the smaller one is 30S.

(*b*) **80S Ribosomes:** Their sedimentation coefficient is 80S and is found in eukaryotes. In 80S ribosomes, the large sub unit is 60S and smaller one is 40S as shown in Fig. 3.9.

According to the most recent account of James A. Lake (1981) the smaller sub units consists of a head, a base and separating from head with the help of a cleft, a platform (Fig. 3.10 A). Similarly, the larger subunit consists of a ridge, a central protuberance and stalk. The ridge and the central protuberance are separated with the help of valley as shown in Fig. 3.10 (A and B). Both parts of ribosomes are attached together with the help of Mg^{2+}. Ribosome plays an important role in protein synthesis.

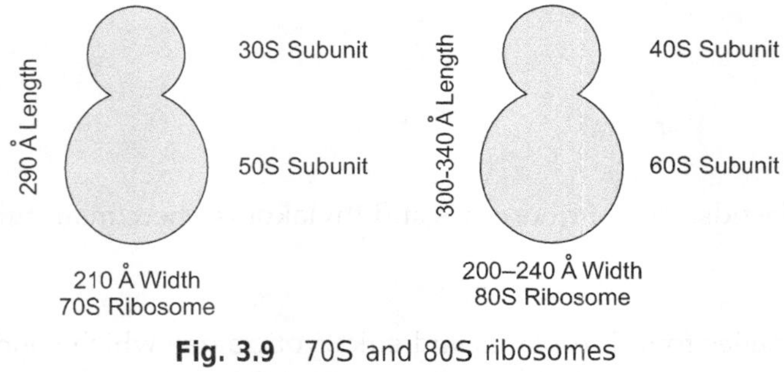

Fig. 3.9 70S and 80S ribosomes

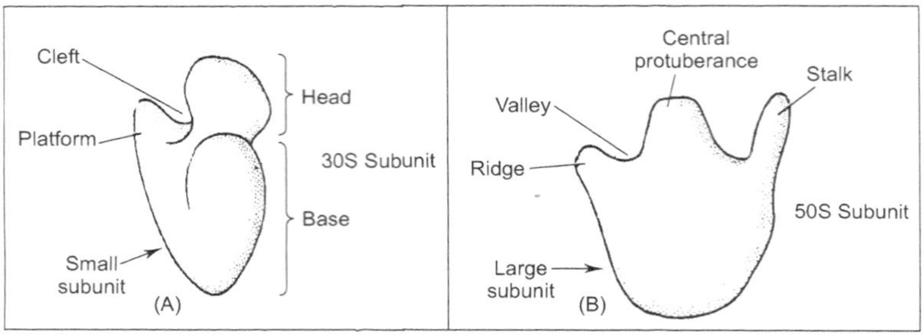

Fig. 3.10 Parts of ribosomes (A and B)

Centriole

A **centriole** in biology is a hollow cylindrical , barrel-shaped organelle found in most animal cells, and cells of fungi and algae though absent in higher plants. The walls of centrioles are composed of nine triplets of microtubules, each arranged at a right angle (so when looking down the tube, one end of the triplet pair points slightly out and the other end points slightly inwards)as shown in Fig. 3.11. In cells, they are usually found in pairs at a right-angle to each other, forming a centrosome. Centrioles are important in the cell division process, organising the spindle upon which the chromosomes are pulled apart. Some animal cells may be able to separate without centrioles. Centrioles assist the cell through the process of mitosis. Centrioles are involved in the organization of the mitotic spindle and in the completion of cytokinesis. Centrioles were previously thought to be required for the formation of a mitotic spindle in animal cells. Centrioles are a very important part of centrosomes, which are involved in organizing microtubules in the cytoplasm. The position of the centriole determines the position of the nucleus and plays a crucial role in the spatial arrangement of the cell.

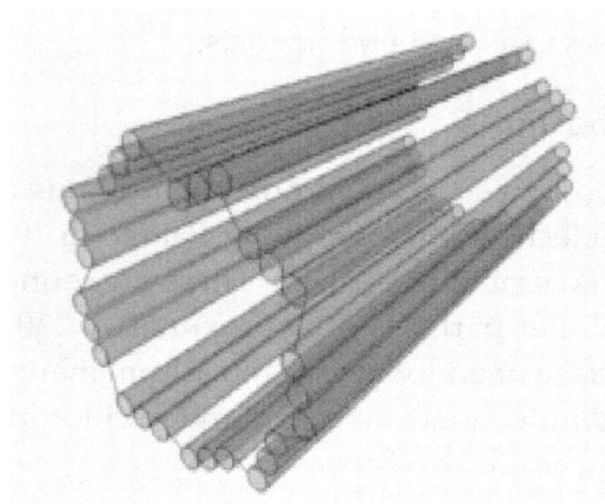

Fig. 3.11 Three-dimensional view of a centriole

Plastids

Plastids are special types of cytoplasmic organelles found in plant cells. They are of three types:

(*a*) Leucoplasts
(*b*) Chloroplasts
(*c*) Chromoplasts

Leucoplasts

They are colourless plastids, lack of ribosomes and thylakoids.There main function is storage.

Chloroplasts

Chloroplasts are organelles found in plants and eukaryotic algae which conduct photosynthesis. The fluid within the chloroplast is called the *stroma*, corresponding to the cytoplasm of the bacterium, and contains tiny circular DNA and ribosomes. Within the stroma are stacks of thylakoids, the sub-organelle where photosynthesis actually takes place. A stack of thylakoids is called a granum (Fig. 3.12). A thylakoid looks like a flattened disk and inside is an empty area called the thylakoid space. The photosynthesis reaction takes place on the surface of the thylakoid. The structure and biochemistry of chloroplasts are also similar to those of Cyanobacteria. Carbon dioxide is converted into carbohydrates via the Calvin cycle, which takes place in the stroma. *Thylakoids*, tiny membrane-enclose sacs, are present and in most forms are arranged into stacks called *grana*. The thylakoid membranes are the site of photosynthesis, converting the energy of light into ATP and oxidizing water to oxygen. The photosynthetic proteins in the membrane bind chlorophyll, which is present with various accessory pigments. These give chloroplasts and their colours, which may be green, golden, brown, or red.

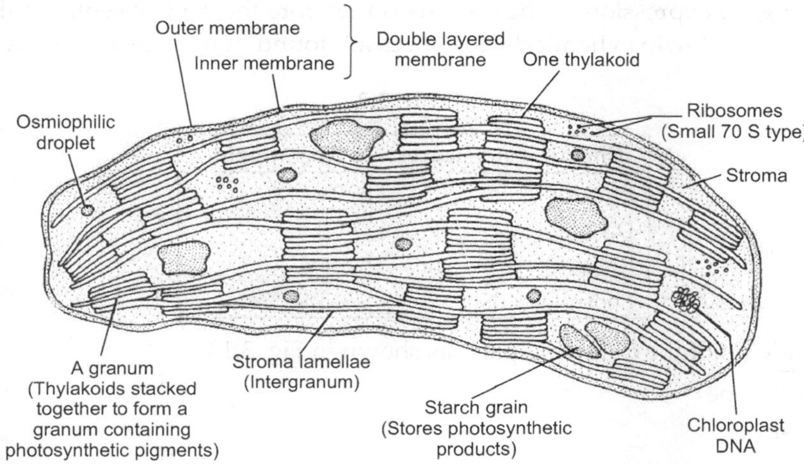

Fig. 3.12 Electron microscopic structure of chloroplast in mesophyll cell of a leaf

Chromoplasts

They are coloured plastids, usually containing yellow (xanthophylls) or orange (carotene) pigments.

Glyoxysomes

They are small, single membrane bound and oval shaped. They consist of an amorphous protein matrix containing enzymes of glyoxylate cycle.

Spherosomes

They are small, single membrane bound organelle associated with synthesis and storage of lipids.

Peroxisomes

They are small, single membrane bound and oval shaped structures of cytoplasm containing oxidizing enzymes in fine granular matrix.

3. Nucleus

The nucleus was first of all reported in 1831 by Scottish botanist Robert Brown in a talk at the Linnean Society of London. Brown was studying orchids microscopically when he observed an opaque area, which he called the areola or nucleus, in the cells of the flower's outer layer. He did not suggest a potential function. In 1838 Matthias Schleiden proposed that the nucleus plays a role in generating cells, thus he introduced the name "Cytoblast" (cell builder). He believed that he had observed new cells assembling around "cytoblasts". Rudolf Virchow (1855) who decisively propagated the new paradigm that cells are generated solely by cells ("Omnis cellula e cellula"). The function of the nucleus remained unclear. In cell biology, the nucleus (plural. nuclei; from Latin nucleus or nuculeus, meaning kernel), also sometimes referred to as the "control center", is a membrane-enclosed organelle found in eukaryotic cells. It contains most of the cell's genetic material, organized as multiple long linear DNA molecules in complex with a large variety of proteins, such as histones, to form chromosomes. The genes within these chromosomes are the cell's nuclear genome. The function of the nucleus is to maintain the integrity of these genes and to control the activities of the cell by regulating gene expression — the nucleus is therefore the control centre of the cell. On the dry weight basis the following chemical substances are found in nucleus as shown in Table: 3.2.

Table 3.2

Chemical substances	Dry weight
Proteins	70%
DNA	10%
RNA	2–3%
Phospholipids	3–5%

Nucleus is made up of the following parts as shown in Fig. 3.13.

- Nuclear envelope
- Nucleoplasm
- Nucleolus
- Chromosomes

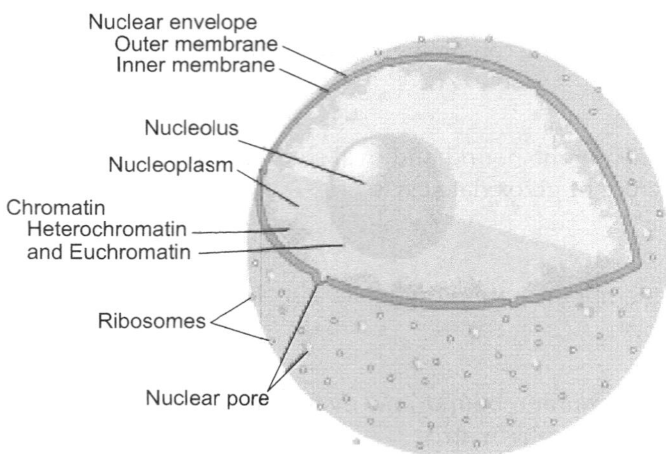

Fig. 3.13 The eukaryotic cell nucleus

Nuclear Envelope and Nuclear Pore

Nucleus is surrounded on all sides' by a membrane called nuclear membrane or nuclear envelope consists of two cellular membranes, an inner and an outer membrane, arranged parallel to one another and separated by 10 to 50 nanometers (nm). The nuclear envelope completely encloses the nucleus and separates the cell's genetic material from the surrounding cytoplasm, serving as a barrier to prevent macromolecules from diffusing freely between the nucleoplasm and the cytoplasm. The inner membrane consists of a fibrous meshwork called nuclear lamina. This nuclear lamina is composed of one or more related proteins called lamins. In addition to providing structural support to the nucleus, the nuclear lamina is thought to serve as a site of chromatin attachment. The outer nuclear membrane is continuous with the membrane of the Rough Endoplasmic Reticulum (RER), and is similarly studded with ribosomes. The space between the membranes is called the perinuclear space and is continuous with the RER lumen as shown in Fig. 3.14. Nuclear pores, which provide aqueous channels through the envelope, are composed of multiple proteins, collectively referred to as nucleoporins. The pores are about 125 million daltons in molecular weight. The pores are 100 nm in total diameter; however, the gap through which molecules freely diffuse is only about 9 nm wide, due to the presence of regulatory systems within the centre of the pore. This size allows the free passage of small water-soluble molecules while preventing larger molecules, such as nucleic acids and larger proteins, from inappropriately entering or exiting the nucleus. These large molecules must be actively transported into the nucleus.

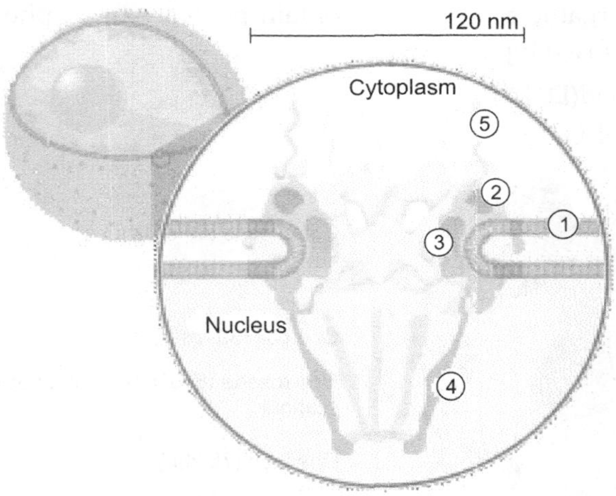

Fig. 3.14 A cross section of a nuclear pore on the surface of the nuclear envelope (1) Other diagram labels show (2) the outer ring, (3) spokes, (4) basket, and (5) filaments

Nucleoplasm

The nucleoplasm is that part of the protoplasm which is enclosed by nuclear membrane. Large number of ribosomes are encoded in it. In addition to ribosomes densely stained thread like structures called chromatin threads and nucleolus are also embedded in it.

Nucleolus

The nucleolus is a discrete densely stained structure found in the nucleus. It is not surrounded by a membrane. It forms around tandem repeats of rDNA, DNA (as discuss in previous chapter) coding for ribosomal RNA (rRNA). These regions are called Nucleolar Organizer Regions (NOR). The main roles of the nucleolus are to synthesize rRNA and assemble ribosomes. The structural cohesion of the nucleolus depends on its activity, as ribosomal assembly in the nucleolus results in the transient association of nucleolar components, facilitating further ribosomal assembly and hence further association. There are four chief components of a nucleolus:

(*a*) ***An amorphous matrix:*** It is heterogeneous and granules and fibrils scattered in it.

(*b*) ***Fibrillar portion:*** It is made up of ribonucleoprotein.

(*c*) ***Granular portion:*** It is made up of protein plus RNA. These are thought to be precursors of ribosomes and sometimes called nucleolar ribosome.

(*d*) ***Chromatin:*** It is of two types, perinucleolar chromatin and intranucleolar chromatin.

Chromosomes

The cell nucleus contains the majority of the cell's genetic material, in the form of multiple linear DNA molecules organized into structures called chromosomes. During most of the cell cycle these are organized in a DNA-protein complex known as chromatin (Fig. 3.15) and during cell division the chromatin can be seen to form the well defined chromosomes familiar from a karyotype. There are two types of chromatin. Euchromatin is the less compact DNA form and contains genes that are frequently expressed by the cell. The other type, heterochromatin, is the more compact form and contains DNA that are infrequently transcribed. Chromatin which is the most important part of nucleus is nucleoprotein, which is made up of nucleic acids* and basic protein histone. Nucleic acids are very complex organic acids. These contain pentose sugar, phosphate and nitrogenous bases. Nucleic acids are of two types:

(*a*) Deoxyribonucleic acid(DNA)

(*b*) Ribonucleic acid(RNA)

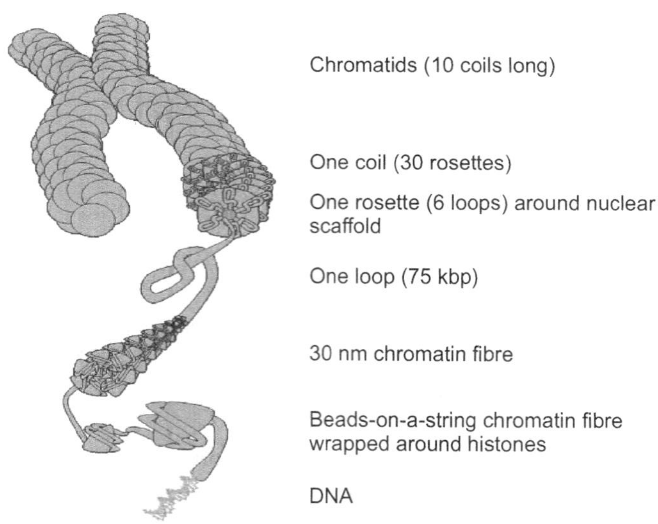

Fig. 3.15 The compacted structure of chromatin

*For more detail about nucleic acid read chapter no.2 biomolecules.

Centrosome

They are found in lower plants, algae and fungi. Centrosome is involved in formation of bipolar spindle during cell division and also in the formation of cilia.

3.1.3 Some Other Types of Cells

Viruses

Viruses are not posses any type of taxonomy. According to International Committee of Taxonomy of viruses, there are no orders, divisions and kingdoms established for viruses. Viruses are not composed of cells, and use anabolic machinery within living host cells to multiply. Viruses lack formal names and most of them acquired their names from sources. Difference between virus and bacteria as shown in following Table 3.3:

Table 3.3 Difference between virus and bacteria

Viruses	Bacteria
Obligate intracellular parasites	Usually free-living, but can be parasites
No ribosomes	Ribosomes
DNA or RNA, not both	DNA and RNA
10–100s of genes	100s–1000s of genes
Tangled phylogeny	Natural phylogeny

Fungi

Fungi are a miscellaneous assemblage of eukaryotes wanting chlorophyll and are filamentous, with the cell wall made up of chitin and coenocytic. They are typically aerobic, found in damp and dark places. They are active producers of various hydrolytic enzymes. The fungi have several unicellular species, such as baker's yeast (Saccharomyces cerevisiae) and fission yeast (Schizosaccharomyces pombe). Some fungi, such as the pathogenic yeast Candida albicans, can undergo phenotypic switching and grow as single cells in some environments and filamentous hyphae in others. Fungi reproduce both asexually, by budding or binary fission, as well by producing spores, which are called conidia when produced asexually, or basidiospores when produced sexually. Fungi are mostly saprophytic with complex life cycles involving spore formation. They are classified into two types: Moulds, Yeasts. Fungi play an important role in ecological cycles, as they rapidly decompose the dead and release carbon and mineral back to the environment.

Algae

Algae are photosynthetic, mostly multicellular, non-motile, branching organism and are eukaryotic. They are more similar to plants. Algae are considered as primary food producers as they convert chemically simple nutrients to complex organic matter. The green algae are a large group of photosynthetic eukaryotes that include many microscopic organisms. Although some green algae are classified as protists, others such as charophyta are classified with embryophyte plants, which are the most familiar group of land plants. Algae can grow as single cells, or in long chains of cells. The green algae include unicellular and colonial flagellates, usually but not always with two flagella per cell, as well as various colonial, coccoid, and filamentous forms. In the Charales, which are the algae most closely related to higher plants, cells differentiate into several distinct tissues within the organism. There are about 6000 species of green algae.

Table 3.4 Difference between animal and plant cell

S.No.	Animal cell	Plant cell
1.	Cell wall absent	Cell wall present, mainly formed of secreted cellulose
2.	Plastids absent	Plastids are present (while absent in bacteria and fungi)
3.	Vacuoles are usually absent	A central large vacuole is present
4.	Division of cytoplasm is by equatorial construction from periphery to centre	Division of cytoplasm starts with the formation of phra-gmoplast, which is made up of microtubule and golgi vesicles, migrating tool equatorial plate
5.	Centriole always present	Centriole absent

Central Dogma of Life

After the victorious fortitude of DNA structure by Watson & Crick in 1953, molecular biologists formulated a broad-spectrum outline of biological inheritance which revealed the following principles as shown in Fig. 3.16 and known as central dogma of life.

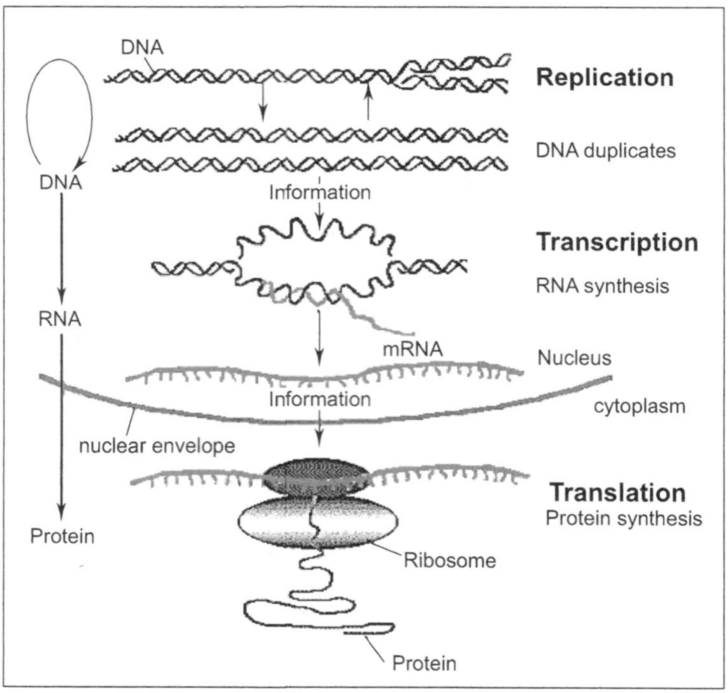

Fig. 3.16 Central dogma depicting steps involved in the expression of protein genes

Table 3.5 Difference between prokaryotic and eukaryotic cells

S.No.	Characteristics	Prokaryotes	Eukaryotes
1.	*Organisms*	Bacteria	Animals, plants, protists, fungi
2.	*Size*	~ 1–10 um	~ 10–100 um
3.	*Nucleus*	Nucleoid region; no real nucleus	Real nucleus with double membrane
4.	*DNA*	Circular (usually)	Linear molecules (chromosomes) with histone proteins
5.	*RNA/protein-synthesis*	Coupled in cytoplasm	RNA-synthesis inside the nucleus protein synthesis in cytoplasm
6.	*Ribosomes*	50S+30S	60S+40S
7.	*Cytoplasmatic structure*	Very few structures	Highly structured by intercellular membranes and cytoskeleton
8.	*Cell movement*	Flagella made of flagellin	Flagella and cilia made of tubulin
9.	*Mitochondria*	None	One to several dozen (though some lack mitochondria)
10.	*Chloroplasts*	None	In algae and plants
11.	*Organization*	Usually single cells	Single cells, colonies, higher organisms with specialized cells
12.	*Cell division*	Binary fission (simple division)	Mitosis (core division) Cytokinesis (cytoplasmatic division)

3.2 MICROBES AND MICROBIOLOGY

A microorganism (from the Greek: mikrós, "small" and organismós, "organism"; also spelled micro organism) or microbe is an organism that is microscopic (too small). The study of microorganisms is called microbiology.

Microorganisms are very diverse; they include bacteria, fungi, archaea and protists; microscopic plants (green algae); and animals such as plankton and the planarian. Some microbiologists also include viruses, but others consider these as non-living. Most microorganisms are unicellular (single-celled), but this is not universal, since some multicellular organisms are microscopic, while some unicellular protists and bacteria, like *Thiomargarita namibiensis*, are macroscopic and visible to the naked eye. Microorganisms live in all parts of the biosphere where there is liquid water, including soil, hot springs, on the ocean floor, high in the atmosphere and deep inside rocks within the Earth's crust. Microbes are also exploited by people in biotechnology, both in conventional food and beverage preparation and in modern technologies based on genetic engineering. However, pathogenic microbes are harmful, since they invade and grow within other organisms, causing diseases. Microorganisms are critical to nutrient recycling in ecosystems as they act as decomposers. As some microorganisms can fix nitrogen, they are a vital part of the nitrogen cycle and recent studies indicate that airborne microbes may play a role in precipitation and weather.

3.2.1 History of Microbiology

The continued existence of microorganisms was hypothesized for many centuries before their authentic discovery in the 17th century. Theories on microorganisms were first made by the Roman scholar Marcus Terentius Varro in a book titled On Agriculture in which he warns against locating a farmstead in the surrounding area of swamps; and because there are bred certain minute creatures which cannot be seen by the eyes, which float in the air and enter the body through the mouth and nose and there cause serious diseases.

In 1546 Girolamo Fracastoro proposed that epidemic diseases were caused by transferable seed like entities that could transmit infection by direct or indirect contact or even without contact over long distances. All these early claims about the existence of microorganisms were speculative in nature and not based on any data or science. Microorganisms were neither confirmed and observed, nor exactly and precisely described until the 17th century. The reason for this was that all these early inquiries lacked the most fundamental tool in order for microbiology and bacteriology to exist as a science and that was the microscope.

Bacteria, and other microorganisms, were first observed by Antonie van Leeuwenhoek in 1676 using a single-lens microscope of his own design. In doing so Leeuwenhoek made one of the most important discoveries in biology and initiated the scientific fields of bacteriology and microbiology. The name "bacterium" was introduced much later, by Ehrenberg in 1828, derived from the Greek meaning "small stick". While Van Leeuwenhoek is frequently cited as the first microbiologist, the first recorded microbiological observation, that of the fruiting bodies of molds, was made earlier in 1665 by Robert Hooke.

The field of bacteriology (later a sub discipline of microbiology) is generally considered to have been founded in the 19th century by Ferdinand Cohn, a botanist whose studies on algae and photosynthetic bacteria led him to describe several bacteria including *Bacillus* and *Beggiatoa*. Cohn was also the first to formulate a scheme for the taxonomic classification of bacteria. Louis Pasteur and Robert Koch were contemporaries of Cohn's and are often considered to be the founders of medical microbiology. Pasteur is most famous for his series of experiments designed to disprove then widely held theory of spontaneous generation, thus solidifying microbiology's uniqueness as a biological science. Pasteur also designed methods for food preservation (pasteurization) and vaccines against a number of diseases such as anthrax, fowl cholera and rabies. Koch is best known for his contributions to the germ theory of disease, proving that specific diseases were caused by specific pathogenic microorganisms. He developed a series of criteria that have become known as the Koch's postulates. Koch was one of the first scientists to focus on the isolation of bacteria in pure culture resulting in his description of several novel bacteria including Mycobacterium tuberculosis, the causative agent of tuberculosis. While Pasteur and Koch are often considered the founders of microbiology, their work did not accurately reflect the true diversity of the microbial world because of their exclusive focus on microorganisms having direct medical relevance. It was not until the late 19th century and the work of Martinus Beijerinck and Sergei Winogradsky, the founders of general microbiology (an older term encompassing aspects of microbial physiology, diversity and ecology), that the true breadth of microbiology was revealed. Beijerinck made two major contributions to microbiology: the discovery of viruses and the development of enrichment culture techniques. While his work on the Tobacco Mosaic Virus established the basic principles of virology, it was his development of enrichment culturing that had the most immediate impact on microbiology by allowing for the cultivation of a wide range of microbes with wildly different physiologies. Winogradsky was the first to develop the concept of chemolithotrophy and to thereby reveal the essential role played by microorganisms in geochemical processes. He was responsible for the first isolation and description of both nitrifying and nitrogen-fixing bacteria.

"Father of Microbiology" Antonie van Leeuwenhoek, the first microbiologist to observe microorganisms using a microscope of his own design.

3.2.2 Classification of Microbes

Classification is a means of bringing order to the mystifying variety of organisms in nature. Once we study the characteristics of an organism we can evaluate it with other organisms to find out similarities and differences. In order to identify and classify microorganisms, we must study their characteristics. It is habitually not possible to study the characteristics of a single microorganism, due to its small size; thus, we study the characteristics of a culture (a population of microorganisms). If we study the characteristics of a culture containing many organisms (usually millions or billions of cells of only one kind), it is as if we are studying the characteristics of a single organism.

Characteristics of Microorganisms

The followings are major characteristics of microorganisms:

- **Morphological characteristics:** Cell shape, size and structure, cell arrangement; occurrence of special structures and development forms; staining reactions; and motility and flagellar arrangement.
- **Chemical composition:** The various chemical constituents of the cells.
- **Cultural characteristics:** Physical conditions and nutritional requirements required for growth.
- **Metabolic characteristics:** The cells carryout chemical reactions and regulate these reactions, and get energy.
- **Genetic characteristics:** characteristics of the hereditary material of the cell, and occurrence and function of other kinds of DNA that may be present, such as plasmid.
- **Antigenic characteristics:** Antigen-antibody reactions, for the rapid identification of particular kinds of microorganisms.
- **Pathogenicity:** The ability to cause disease in various animals and plants or even other microorganisms.
- **Ecological characteristics:** Habitat and the distribution of the organism in nature and the interactions between and among species in natural environment.

Microorganisms can be found almost anywhere in the taxonomic organization of life on the planet. Bacteria and archaea are almost always microscopic, while a number of eukaryotes are also microscopic, including most protists, some fungi, as well as some animals and plants. Viruses are generally regarded as not living and therefore are not microbes, although the field of microbiology also encompasses the study of viruses. To help individuals understand the diverse types of microbes, they are grouped or classified in various ways. Microbes are very diverse and represent all the great kingdoms of life. In fact, in terms of numbers, most of the diversity of life on Earth is represented by microbes. Here is a sketch of the major groups of microorganisms:

Prokaryotes

Prokaryotes are organisms that lack a cell nucleus and the other membrane bound organelles. They are almost always unicellular. The prokaryotes are the most diverse and abundant group of organisms on Earth and inhabit practically all environments where some liquid water is available and the temperature is below + 140°C. They are found in sea water, soil, air, animals' gastrointestinal tracts, hot springs and even deep beneath the Earth's crust in rocks. Practically all surfaces which have not been specially sterilized are covered by prokaryotes. The number of prokaryotes on Earth is estimated to be around five million trillion trillion, or 5 × 1030, accounting for at least half the biomass on Earth. Prokaryotes are consisting of two domains, archaea and bacteria.

- Archaea
- Bacteria

Archaea

Archaea are also single-celled organisms that lack nuclei. In the past, the differences between bacteria and archaea were not recognised and archaea were classified with bacteria as part of the kingdom Monera. However, in 1990 the microbiologist Carl Woese proposed the three-domain system that divided living things into bacteria, archaea and eukaryotes. Archaea differ from bacteria in both their genetics and biochemistry. For example, while bacterial cell membranes are made from phosphoglycerides with ester bonds, archaean membranes are made of ether lipids. Archaea were originally inhabited in extreme environments, such as hot springs, but have since been found in all types of habitats. These organisms are also common in soil and play a vital role in ammonia oxidation.

Bacteria

Bacteria are practically all invisible to the naked eye, with a few extremely rare exceptions, such as Thiomargarita namibiensis. They lack membrane-bound organelles, and can function and reproduce as individual cells, but often aggregate in multicellular colonies. Their genome is usually a single loop of DNA, although they can also harbour small pieces of DNA called plasmids. These plasmids can be transferred between cells through bacterial conjugation. Bacteria are surrounded by a cell wall, which provides strength and rigidity to their cells. They reproduce by binary fission or sometimes by budding, but do not undergo sexual reproduction. Some species form extraordinarily resilient spores, but for bacteria this is a mechanism for survival, not reproduction. Under optimal conditions bacteria can grow extremely rapidly and can double as quickly as every 10 minutes.

Eukaryotes

Unicellular eukaryotes are those eukaryotic organisms that consist of a single cell throughout their life cycle. This qualification is significant since most multicellular eukaryotes consist of a single cell called a zygote at the beginning of their life cycles. Microbial eukaryotes can be either haploid or diploid and some organisms have multiple cell nuclei. However, not all microorganisms are unicellular as some microscopic eukaryotes are made from multiple cells.

- Algae (as discussed in point 3.1.3)
- Fungi (as discussed in point 3.1.3)
- Protozoa
- Animals

Protozoa

The protists are most commonly unicellular and microscopic. This is a highly diverse group of organisms that are not easy to classify. Several algae species are multicellular protists, and slime molds have unique life cycles that involve switching between unicellular, colonial, and multicellular forms. The number of species of protozoa is uncertain, since we may have identified only a small proportion of the diversity in this group of organisms.

Animals

Mostly animals are multicellular, but some are too small to be seen by the naked eye. Microscopic arthropods include dust mites and spider mites. Microscopic crustaceans include copepods and the cladocera, while many nematodes are too small to be seen with the naked eye. Another particularly common group of microscopic animals are the rotifers, which are filter feeders that are usually found in fresh water. Micro-animals reproduce both sexually and asexually and may reach new habitats as eggs that survive harsh environments that would kill the adult animal. However, some simple animals, such as rotifers and nematodes, can dry out completely and remain dormant for long periods of time.

Viruses

According to International Committee of Taxonomy of viruses, there are no orders, divisions and kingdoms established for viruses. (As discussed in point 3.1.3)

3.2.3 Applications of Microbes

Several microbes are responsible for numerous valuable processes such as industrial fermentation (*e.g.*, the production of alcohol, vinegar and dairy products), antibiotic production and as vehicles for cloning in higher organisms. Scientists have also exploited their knowledge of microbes to produce biotechnologically important enzymes such as Taq polymerase, reporter genes for use in other genetic systems and novel molecular biology techniques such as the yeast two-hybrid system. In Table 3.6 listed some of the major products of bioprocessing.

Bacteria can be used for the industrial production of amino acids. *Corynebacterium glutamicum* is one of the most important bacterial species with an annual production of more than two million tons of amino acids, mainly L-lysine and L-glutamate. A variety of biopolymers, such as polysaccharides, polyesters, and polyamides, are produced by microorganisms. Microorganisms are used for the biotechnological production of biopolymers with tailored properties suitable for high-value medical application such as tissue engineering and drug delivery. Microorganisms are used for the biosynthesis of cellulose, poly (gamma-glutamic acid), xanthan, alginate, cyanophycin, levan, hyaluronic acid, organic acids, polysaccharide, oligosaccharides and polyhydroxyalkanoates.

Microorganisms are beneficial for microbial biodegradation or bioremediation of domestic, agricultural and industrial wastes and subsurface pollution in soils, sediments and marine environments. The ability of each microorganism to degrade toxic waste depends on the nature of each contaminant. Since sites typically have multiple pollutant types, the most effective approach to microbial biodegradation is to use a mixture of bacterial species and strains, each specific to the biodegradation of one or more types of contaminants.

There are also various claims concerning the contributions to human and animal health by consuming probiotics (bacteria potentially beneficial to the digestive system) and/or prebiotics (substances consumed to promote the growth of probiotic microorganisms). Recent research has suggested that microorganisms could be useful in the treatment of cancer. Various strains of non-pathogenic clostridia can infiltrate and replicate within solid tumours. Clostridial vectors can be safely administered and their potential to deliver therapeutic proteins has been demonstrated in a variety of preclinical models.

Table 3.6 Major products of biological processing

Fermentation product	Typical organism used	Approximate world market size (kg. per year)
Bulk organics		
Ethanol (non-beverage)	Saccharomyces cerevisiae	2×10^{10}
Acetone/butanol	Clostridium acetobutylicum	2×10^{6} (butanol)
Biomass		
Starter cultures and yeasts for food and agriculture	Lactic acid bacteria or Bakers' yeast	5×10^{8}
Single-cell protein	Pseudomonas methylotrophus or Candida utilis	$0.5–1 \times 10^{8}$
Organic acids		
Citric acid	Aspergillus niger	$2–3 \times 10^{8}$
Gluconic acid	Aspergillus niger	5×10^{7}
Lactic acid	Lactobacillus delbrueckii	2×10^{7}
Itaconic acid	Aspergillus itaconicus	
Amino acids		
L-Glutamic acid	Corynebacterium glutamicum	3×10^{8}
L-Lysine	Brevibacterium flavum	3×10^{7}
L-Phenylalanine	Corynebacterium glutamicum	2×10^{6}
L-Arginine	Brevibacterium flavum	2×10^{6}
Others	Corynebacterium spp.	1×10^{6}

Microbial transformations		
Steroids	Rhizopus arrhizus	
D-sorbitol to L-sorbitol (in Vitamin C production)	Acetobacter suboxydans	4×10^7
Antibiotics		
Penicillins	Penicillium chrysogenum	$3–4 \times 10^7$
Cephalosporins	Cephalosporium acremonium	1×10^7
Tetracyclines (*e.g.*, 7-chlortetracycline)	Streptomyces aureofaciens	1×10^7
Macrolide antibiotics (*e.g.*, erythromycin)	Streptomyces erythreus	2×10^6
Polypeptide antibiotics (*e.g.*, gramicidin)	Bacillus brevis	1×10^6
Aminoglycoside antibiotics (*e.g.*, streptomycin)	Streptomyces griseus	
Aromatic antibiotics (*e.g.*, griseofulvin)	Penicillium griseofulvum	
Extracellular polysaccharides		
Xanthan gum	Xanthomonas campestris	5×10^6
Dextran	Leuconostoc mesenteroides	Small
Nucleotides		
5′-Guanosine monophosphate	Brevibacterium ammoniagenes	1×10^5
Enzymes		
Proteases	Bacillus spp.	6×10^5
α-Amylase	Bacillus amyloliquefaciens	4×10^5
Glucoamylase	Aspergillus niger	4×10^5
Glucose isomerase	Bacillus coagulans	1×10^4
Pectinase	Aspergillus niger	1×10^4
Rennin	Mucor miehei or recombinant yeast	1×10^4
All others		5×10^4
Vitamins		
B_{12}	Propionibacterium shermanii or Pseudomonas denitrificans	1×10^4
Riboflavin	Eremothecium ashbyii	
Ergot alkaloids	Claviceps paspali	5×10^3
Pigments		
Shikonin	Lithospermum erythrorhizon (Plant-cell culture)	60
β-Carotene	Blakeslea trispora	
Vaccines		
Diphtheria	Corynebacterium diphtheriae	< 50

Tetanus	Clostridium tetani	
Pertussis (whooping cough)	Bordetella pertussis	
Poliomyelitis virus	Live attenuated viruses grown in monkey kidney or human diploid cells	
Rubella	Live attenuated viruses grown in baby-hamster kidney cells	
Hepatitis B	Surface antigen expressed in recombinant yeast	
Therapeutic proteins		< 20
Insulin	Recombinant Escherichia coli	
Growth hormone	Recombinant Escherichia coli or recombinant mammalian cells	
Erythropoietin	Recombinant mammalian cells	
Factor VIII-C	Recombinant mammalian cells	
Tissue plasminogen activator	Recombinant mammalian cells	
Interferon α_2	Recombinant Escherichia coli	
Monoclonal antibodies	Hybridoma cells	< 20
Insecticides		
Bacterial spores	Bacillus thuringiensis	
Fungal spores	Hirsutella thompsonii	

REVIEW QUESTIONS

1. What is cell? Explain the brief history of cell.
2. Give the difference between prokaryotic and eukaryotic cell.
3. Give the difference between plant cell and animal cell.
4. Explain different types of cell organelles of eukaryotic cell.
5. Explain the central dogma of molecular cell biology.
6. Elucidate the difference between bacteria and virus.
7. Explain in detail the cell nucleus.
8. Explain the various functions of cell organelles.
9. Give the brief history of microbiology.
10. Give the classification of microbes.
11. Give the applications of microbes in human welfare.

4

BIOINFORMATICS

4.1 INTRODUCTION OF BIOINFORMATICS

Bioinformatics is the discipline of quantitative analysis of information relating to biological macromolecules with the help of computers. We use computer programs to make inferences from the data archives of modern molecular biology, to make connections amid them and to get useful and interesting predictions.

Fredij Tekaia proposed the definition of bioinformatics as *'the mathematically, statistically and computing methods that aim to solve biological problems using DNA and amino acid sequences and related information'*.

This emerging field is turning out to be a well opted carrier choice of the twenty first century.

The term bioinformatics was coined by Paulien Hogeweg in 1979 for the study of informatics processes in biotic systems. The main use of bioinformatics is in genomics and genetics, particularly in those areas of genomics involving large-scale DNA sequencing. Bioinformatics necessitate the making and development of databases, computational and statistical techniques, algorithms, and theory to resolve formal and practical problems arising from the biological data analysis and management. Over the past few decades hasty developments in genomic and other molecular research technologies and developments in information technologies have united to turn out a remarkable amount of information related to molecular biology. Widespread performance bioinformatics include mapping and analyzing DNA and protein sequences, aligning different DNA and protein sequences to compare them and creating and viewing 3-D models of protein structures.

The primary goal of bioinformatics is to better understand about living cells and how it functions at the molecular level. By analysing raw molecular sequence and structural data, bioinformatics research can generate new experience and provide a global insight of the cell. The cause that the functions of a cell can be well understood by analysing sequence data is finally because the flow of genetic information is dictated by the 'central dogma' of biology in which DNA is transcribed to RNA, which is further translated to proteins. Cellular functions are primarily performed by proteins whose capabilities are ultimately determined by their sequences. Though, solving functional problems using sequence and sometimes structural approaches has proved to be a productive venture. Bioinformatics is also focus on developing and applying computationally intensive techniques such as pattern recognition, data mining, machine learning algorithms, and visualization. most important research attempts in the field include sequence alignment, gene finding, genome assembly, drug design, drug discovery, protein structure alignment, protein structure prediction, prediction of gene expression and protein-protein interactions, genome-wide association studies.

Scope of Bioinformatics

Bioinformatics derives acquaintance from computer analysis of biological data. These can consists of the information stored in the genetic code, but can also experimental results from various sources. Bioinformatics is a rapidly developing branch of biology and is highly interdisciplinary, using technique and concept from informatics, statistics, mathematics, chemistry, biochemistry, physics and linguistics. Bioinformatics integrates the advances in the area of computer science, information science and information technology to solve the problem in life sciences. The core data comprise of the genomes and proteomes of human and other organisms, 3D molecular structures and functions of proteins, microarray data, metabolic pathways, cell lines, and hybridoma technology, biodiversity. Internet is utilized to collect data, distribute and access data, which is later analysed with statistics and mathematics based tools. Bioinformatics has a key role to play in the areas such as functional genomics, proteomics, protein engineering, metabolic pathways engineering, Pharmacogenomics, discovery of new drugs and vaccines, agro- biotechnology etc.

4.2 APPLICATIONS OF BIOINFORMATICS

With the help of bioinformatics, it is possible to investigate a various new areas in biology, some are given below:

Comparative Genomics

By comparing the genes of various organisms, it is feasible to trace evolutionary pathways by which one organism could have evolved in to another. Such studies can not only fling new light on advancement, but also provide indication for the migration of species, thus bringing new evidence to historical and anthropological studies.

Microbial Genomics

The genomes of bacteria can assist throw light on energy sources, environmental monitoring to identify pollutants, discover disease-producing properties of genes and get better industrial efficiency. For these reasons, in 1994, The U.S. Department of Energy initiated (MGP) microbial genome projects to sequence the genome of bacteria useful in energy production, environmental clean-up, industrial processing, and toxic waste production.

Molecular Medicine

More emphasis needs to be given on tracing the fundamental cause of diseases rather than treating symptoms. Molecular medicine is the treatment of injury or disease at the molecular level. Study of molecules as they related to health and disease and manipulating those molecules to improve the diagnosis, prevention, and treatment of disease. The completion of the human genome means that we can search for the genes directly associated with different diseases and begin to understand the molecular basis of these diseases more clearly.

Drug Development

With an improved understanding of disease mechanism and using computational tools to indentify and validate new drug targets, more specific medicines that act on cause, not merely the symptoms, of the disease can be developed . These highly specific drugs promise to have fewer side effects them many of today's medicines.

Gene Therapy

Gene therapy is the approach used to treat, cure or even prevent disease by changing the expression of a person's gene .The crucial first step in gene therapy is the introduction of the DNA sequences of interest into the appropriate target cells. Currently this field is in its infantile stage and has more possibility.

Biodiversity

Bioinformatics provides a genetic measure of biodiversity in addition to species and physical measure. Eventually, it is the genetic diversity in a biosphere that will provide the correct measure of its extent.

DNA Forensics

DNA profile of individual, called fingerprints can help in identifying criminals, establishing family relationships, protecting wild life species and matching organ donors.

Risk Assessment

More research on human genome can help to assess individual risk on exposure to toxic elements as resistance to external agents varies from person to person.

Analysis of Gene Expression

A physical or chemical change in a living system is not caused by a single gene but by the combined effect of many genes. Understanding the action of many gens on a single condition will, one day, provide a genetic basis for disease and change control.

Analysis of Gene Regulation

Regulation is the chain of events, starting with an extracellular event and leading to a change in the activity of proteins.

Bioarchaeology, Anthropology, Evolution, and Human Migration

Understanding the human and other genome will help to understand evolution inheritance, traits and disease carriers. The study of the genome comparison across organisms can help to understand similar genes with associated disease.

Agriculture, Livestock Breeding, and Bioprocessing

Genome explore on plants can provide nutritious, disease-resistant pesticide free crop. Even alternate use of crops can be found, *e.g.*, tobacco has been found to produce bacterial enzymes that break down some explosive like TNT and dinitroglycerin.

Pharmacogenomics

Pharmacogenomics holds the promise that drugs might one day be custom-made for humans. Environment, diet, age, lifestyle, and state of health all can influence a person's response to medicines, but understanding an individual's genetic makeup is thought to be the key to create personalized drugs with greater efficacy and safety.

Analysis of Protein Expression

Protein microarrays and High Throughput (HT) Mass Spectrometry (MS) can provide a picture of the proteins present in a biological sample. Bioinformatics is very much concerned in making sense of protein microarray and HT MS data; the former approach faces similar problems as with microarrays targeted at mRNA, the latter involves the problem of matching large amounts of mass data against predicted masses from protein sequence databases, and the complicated statistical analysis of samples where multiple, but incomplete peptides from each protein are detected.

Analysis of Mutations in Cancer

The genomes of affected cells (cancerous cells), are rearranged in complex or even unpredictable ways. Enormous sequencing efforts are used to recognize earlier unknown point mutations in a variety of genes in cancer. Bioinformaticians continue to make specialized automated systems to deal with the sheer volume of sequence data produced, and they create new algorithms and software to compare the sequencing results to the growing collection of human genome sequences and germ line polymorphisms. New physical detection technology are employed, such as oligonucleotide microarrays to identify chromosomal gains and losses (called comparative genomic hybridization), and single-nucleotide polymorphism arrays to detect known point mutations. These detection methods simultaneously measure several hundred thousand sites throughout the genome, and when used in high-throughput to measure thousands of samples, generate terabytes of data per experiment. Again the new types of data in huge amounts produce new opportunities for bioinformaticians. The data is often found to have significant variability, or noise, and thus Hidden Markov model and change-point analysis methods are being developed to infer real copy number changes. Another type of data that requires novel informatics development is the analysis of damage found to be persistent amid several tumours.

Genome Annotation

In the view of genomics, annotation is the procedure of marking the genes and other biological characteristics in a DNA sequence. The first genome annotation software system was designed by Dr. Owen White in 1995 at 'The Institute for Genomic Research'; his team have sequenced and analyzed the first genome of a free-living organism of the bacterium Haemophilus influenzae. Dr. White built a software system to find the genes (places in the DNA sequence that encode a protein), the transfer RNA, and other features, and to make initial assignments of function to those genes. Most recent genome annotation systems work similarly, but the programs available for analysis of genomic DNA are frequently shifting and recovering.

4.3 NATIONAL CENTRE FOR BIOTECHNOLOGY INFORMATION (NCBI)

NCBI was established in the year 1988, as a part of the National Library of Medicine at the National Institute of Health, Bethesda, Maryland, USA. Its aim was to produce public databases, develop software tools for sequence analysis, and disseminate biomedical information, primarily to assist the research in computational biology. The role of NCBI is to maintain various biological databases, important amid them in GenBank; the nucleic acid sequence database, to which data are submitted by researchers from all over the world. All the databases of NCBI are grouped into primary and derivative databases. Primary databases contain original biological data. Their content controlled by the submitter. Examples of primary databases are SNP, GEO (Gene Expression Omnibus), GenBank. Derivative databases are built up from primary data and content controlled by third party (NCBI). Examples of derivative databases are Refseq (Reference Sequence), RefSNP, UniGene, NCBI Protein, GEO datasets, TPA (Third Party Annotation), Structure, and Conservation Domain Database (CDD). NCBI provides data retrieval systems i.e. Entrez; it also provide computational resources for the analysis of GenBank data and a range of other biological data.

4.4 BIOLOGICAL DATABASES

A biological database is a large, organized body of persistent data, usually associated with computerized software designed to update, query, and retrieve components of the data stored within the system. A simple database might be a single file containing many records, each of which includes the same set of information. There are many specialized databases that have mushroomed over an incredibly short period of time. The examples are the nucleotide sequences, protein sequences, and the 3D structural data produced by x-ray crystallography, NMR, and microarrays. Most of the databases in the field of bioinformatics are freely accessible from the internet. Biological databases are libraries of life sciences information, collected from scientific experiments, published literature, high-throughput experiment technology, and computational analyses. They contain information from research areas including genomics, proteomics, metabolomics, microarray gene expression, and phylogenetics. Information contained in biological databases includes gene function, structure, localization (both cellular and chromosomal), clinical effects of mutations as well as similarities of biological sequences and structures. Relational concepts of computer science and Information retrieval concepts of digital libraries are important for understanding biological databases. Biological database design, development, and long-term management are a core area of the discipline of bioinformatics. Data contents include gene sequences, textual descriptions, attributes and ontology classifications, citations, and tabular data. These are often described as semi-structured data, and can be represented as tables, key delimited records, and XML structures. Biological databases are an important tool in assisting scientists to understand and explain a host of biological phenomena from the structure of biomolecules and their interaction, to the whole metabolism of organisms and to understanding the evolution of species. This knowledge helps make possible the fight against diseases, assists in the development of medications and in discovering basic relationships amongst species in the history of life. Three types of database structures are used in existing biological databases; flat files, relational, and object oriented. Biological databases can be generally divided in to three categories: primary database, secondary database, and specialized databases. Primary databases enclose original biological data. They are records of raw sequence or structural data submitted by the particular community. Protein Data Bank (PDB) and GenBank are examples of primary databases. Secondary databases contain computationally processes or manually curated information, based on original information from primary databases. Translated protein sequence database containing functional annotation belong to this category. SWISS-Prot and Protein Information Resources (PIR) are the example of this. Specialized database are those that provide to a particular research interest; e.g. Flybase, HIV sequence database and Ribosomal Database Project are databases that specialize in a particular organism or particular type of data.

Categories of Databases on the Basis of Data

On the basis of the data content, databases are categorised in the following way:

- Taxonomy databases
- Genome databases
  - (*a*) Genomics databases
  - (*b*) Human and other vertebrate genomes
- Sequence databases
  - (*a*) Nucleotide sequence databases
  - (*b*) Protein sequence databases
  - (*c*) RNA sequence databases
- Structure databases
- Proteomic databases
- Microarray databases
- Chemical databases
- Expression databases
- Enzyme databases
- Pathway databases
- Diseases databases
- Literature databases
- Other molecular biology databases

Here, we emphasize only on the Nucleotide sequence databases, Protein sequence databases and structural databases.

4.4.1 Nucleotide Sequence Databases

These are data repositories that accept nucleic acid sequence data and make it freely available to the public. The data in these repositories is heterogeneous with respect to the source of material, quality, annotation and the intended completeness of the sequence relative to its biological target. Much of the relevant and useful information available today is housed in secondary databases. These databases have information *i.e.*, either curated on annotated. DDBJ, EMBL and TrEMBL are examples of primary databases, and Swiss-Prot, Prosite and PDB are example of secondary databases. Databases EMBL, GenBank, and DDBJ are the three primary nucleotide sequence databases. These databases include sequences submitted directly by scientist and genome sequencing groups and sequences taken from literature and patents. These three databases exchange and update data on a daily bases to achieve optimal synchronization. The data is exchanged using a common formate for data distribution. DDBJ, EMBL and GeneBank hold to documented guidelines to regulate the content and syntax of the database entries. The International Nucleotide Sequence Databases Collaboration consists of nucleotide sequence databases from three groups. They are:

- DNA Databank of Japan(DDBJ) of National Institute of Genetics, Japan
- GenBank of National Centre for Biotechnology Information(NCBI), USA
- The European Molecular Biology Labs (EMBL) Data Library of European Bioinformatics Institute, UK

EMBL

The European Molecular Biology Laboratory (EMBL) is a nucleotide sequence database maintained by the Data Library of European Bioinformatics Institute, UK (EBI). The current databases release (Release 81, Dec. 2004), complete with release notes and user manual, are available from the EBI servers. EMBL database is a part of the International Nucleotide Sequence Database Collaboration, which includes GenBank (USA) and DDBJ (Japan). The intend of this; is to collect and present nucleotide sequence and annotation with complete global coverage. The main purpose of EMBL Nucleotide Sequence Database is to build, maintain, and prepare biological databases and other computational services to support data deposition and data analysis, and make them accessible to the scientific community. It is a gigantic storehouse of biological data and bio-software.

GenBank

The GenBank nucleotide database is maintained by National Center for Biotechnology Information (NCBI), USA, which is a part of the National Institute of Health (USA). GenBank is the primary sequenced database and searched through the Entrez gateway at NCBI. It can also be downloaded as entire database as flat files. It is a complete public database of nucleotide sequences, supporting bibliographic and biological annotation. GenBank makes data accessible at no cost over the internet, via FTP and a broad range of web-based retrieval and analysis services which drive on the GenBank data. There are two ways to search for sequences in GenBank. One is using text based keywords similar to PubMed search. The other is using molecular sequences to search by sequence similarity using BLAST. Gene bank is built mainly from the submission of sequence data from authors and from the bulk submission of Expressed Sequence Tag (EST), Genome Survey Sequence (GSS), and other high throughput data from the sequence centres. GenBank, along with the EMBL data library in Europe and the DNA Databanks of Japan (DDBJ) comprises the International Nucleotide Sequence Databases (INSD). It is a mutual approach for exchanging data every day to make sure a uniform and comprehensive collection of sequence information (Figure 4.1).

DDBJ

DNA Databank of Japan (DDBJ), maintained and run by National Institute of Genetics (NIG) (http: // www.nig.ac.jp), Japan, collects and supplies DNA data. This data was established in 1986. DDBJ has been functioning as the International Nucleotide Sequence Databases (INSD) in collaboration with EBI/ EMBL and NCBI/ GenBank. The DDBJ release 60 was completed in Dec.2004. Year 2005 an increment of 50% in the amount of data stored at DDBJ. Among them, the genome data of human, ascidians, and rice are the top three areas where data submissions have seen an increase. One can search for entries by the accession number, FASTA/BLAST, keywords, and regular expressions. Getentry, as a resource at DDBJ; it provides an easy way to retrieve entries from the various databases at DDJB. SAKURA is a nucleotide sequence data submission system through the World Wide Web server at DDBJ and it has been functional since 1995. Users can enter and submit nucleotide and translated amino acid sequence through this system.

4.4.2 Protein Sequence Databases

There are two categories of universal protein databases: one that stores simple archives of sequence data, and the other that stores annotated databases. In the annotated databases, additional information is added to the sequence record. Databases act as global databases having entries from all the species and specialized data about specific families or groups of proteins, or about the proteins of a specific organism.

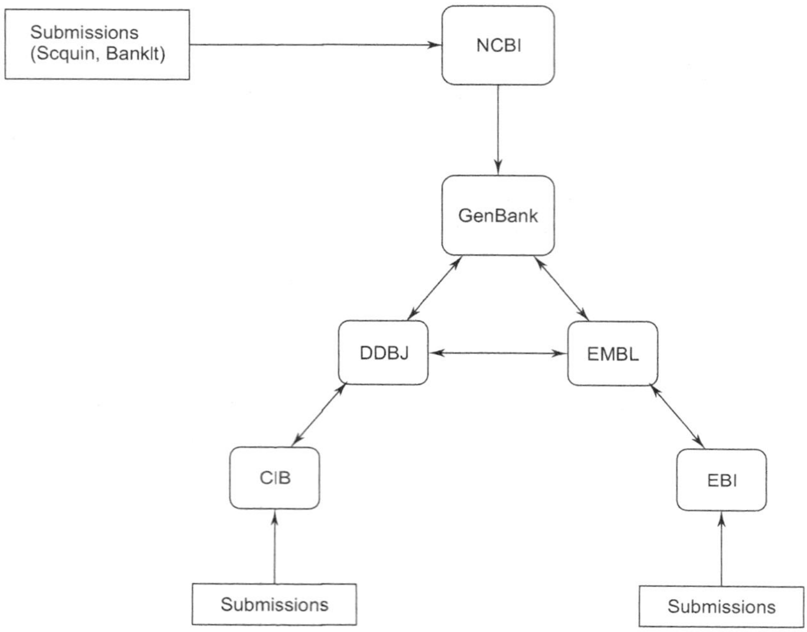

Fig. 4.1 International Sequence Database Collaboration

The UniProt Consortium

UniProt is a central repository of protein sequence and function. It is enriched by information shared from those contained in Swiss-Prot, TrEMBL , PIR and many more sources. This makes it the world's most comprehensive resource on protein information. The UniProt Consortium comprises the European Bioinformatics Institute (EBI), the Swiss Institute of Bioinformatics (SIB), and the Protein Information Resource (PIR). EBI located at the Wellcome Trust Genome Campus in Hinxton, UK, hosts a large resource of bioinformatics databases and services. SIB, located in Geneva, Switzerland, maintains the ExPASy (Expert Protein Analysis System) servers that are a central resource for proteomics tools and databases. PIR, hosted by the National Biomedical Research Foundation (NBRF) at the Georgetown University Medical Center in Washington, DC, USA, is heir to the oldest protein sequence database, Margaret Dayhoff's Atlas of Protein Sequence and Structure. In 2002, EBI, SIB, and PIR joined forces as the UniProt consortium.

The UniProt databases consist of the following database layers:

- The UniProt Knowledgebase (UniProtKB)
- The UniProt Reference Clusters (UniRef)
- The UniProt Archive (UniParc)
- The UniProt Metagenomics and Environmental Sequences database (UniMES)

The UniProt Knowledgebase (UniProtKB) caters the central database of protein sequences with annotation and functional information. PIR-SID sequences misplaced from SWISS-Prot and TrEMBL were incorporated in to UniProt. UniProt Knowledgebase has two parts. One part contains manually annotated records, and is referred to as 'UniProt/ Swiss-Prot. The other part contains computationally analysed records, which have to be manually annotated, and is referred to as UniProt/ TrEMBL. The knowledge base aims to describe in a single record all the protein products derived from a certain gene from a certain species and gives not only the entire record an accession number, but also assigns alternative splicing, proteolytic cleavage, and post translational modification isoform identifiers, to each form of the derived proteins.

The UniProt Reference (UniRef) databases provide non redundant data collections based on the UniProt Knowledgebase and UniProt archive to obtain full coverage of sequence space at several resolutions.

The UniProt archive (UniParc) provides a stable, comprehensive, non redundant sequence collection by storing the complete body of publicly available protein sequence data. The basic information stored with each UniParc entry are the identifier, the sequence, cyclic redundancy check number, source databases with accession and version numbers and a time stamp. Other information can be retrieved from the source databases. Every source database accession number is tagged with its status in that database. The UniProt Metagenomics and Environmental Sequences database (UniMES) is a repository exclusively developed for metagenomic and environmental sequence data.

Swiss-Prot

Swiss-Prot is an annotated biological database of protein sequences maintained by the Swiss Institute of Bioinformatics (SIB), Switzerland and the European Bioinformatics Institute (EBI). It was created in 1986 by Amos Bairoch during his PhD. Swiss-Prot strives to provide reliable protein sequences associated with a high level of annotation (such as the description of the function of a protein, its domains structure, post-translational modifications, variants, etc.), a minimal level of redundancy and high level of integration with other databases.

TrEMBL

TrEMBL, Translation of EMBL nucleotide sequence database was introduced in 1996. It is computer annotated supplementary database to Swiss-Prot. There has been an amplify in the amount of data flow from the genome projects to the sequence databases. Therefore, there are challenges to provide quick database annotation. Although it is required to maintain the high quality in annotation, it is also essential to make sequences available as swiftly as possible. Computer translation is not entirely perfect, so proteins predicted by the TrEMBL database can be hypothetical, and many TrEMBL entries are poorly annotated. In contrast to Swiss-Prot, which contains only proteins really found in the wild, and Protein Information Resource, which is completely unchecked. It is divided in to two sections:

1. SP- TrEMBL: It contains sequences, which will be eventually incorporated in to the Swiss-Prot.
2. REM-TrEMBL: It contains those sequences which will not be incorporated in to the Swiss-Prot.

PIR-PSD

The Protein Information Resource (PIR) is a division of the National Biomedical Research Foundation (NBRF) in the US. PIR-PSD (Protein sequence database) is a function annotated database of PIR, MIPS, and JIPID at NBRF, USA. This database evolved from the original NBRF protein sequence database developed by Margaret Dayhoff, 1965, which was published as the Atlas of Protein Sequence and Structure. It is the source of annotated protein databases and analysis tools for the researchers. It is usually believed that it does not reach the level of completeness in the entry annotation as much as does Swiss-Prot. While Swiss-Prot and PIR overlap extensively, there are still many sequences which can be found in only one of them. Like Swiss-Prot, this database can be searched for entries or sequence similarity searches. The PIR-PSD flat files are available in the XML format. The PSD sequence file is distributed in the FASTA format. The database can also be downloaded as a set of files.

4.4.3 Structure Databases

The structural databases are the repository of 3-D structure of macromolecules. These databases are very useful for protein structure prediction. In biology, a protein structure database is a database that is modelled around the various experimentally determined protein structures. The aim of most protein structure databases is to organize and annotate the protein structures, providing the biological community access to the experimental data in a useful way. The Protein Data Bank is the primary database used today for 3D structure of biological macromolecules. SCOP (Structural Classification of Proteins) is another database that classifies protein 3Dstructure in a hierarchical design of structural classes. SCOP based on the primary database of PDB. CATH (class, architecture, topology, homologous super family) is another type of databases. MMDB is a structure database associated with the NCBI website.

Protein Data Bank (PDB)

The Protein Data Bank (PDB) was established in 1971 by research collaborator for Structural Bioinformatics (RCSB), Brookhaven National Laboratory, USA. It is the central archive of all experimentally determined protein structure data. Nowadays the PDB is maintained by an international consortium collectively known as the Worldwide Protein Data Bank (wwPDB); it was established in 2003. The mission of the wwPDB is to maintain a single archive of macromolecular structural data that is freely and publicly available to the global community. The Protein Data Bank (PDB) is a repository for the 3-D structural data of bio-macromolecules, such as proteins and nucleic acids. The data has been determined by X-ray crystallography and NMR spectroscopy. The data in this bank contains details of the atomic coordinates, some structural details, and additional data computed from the structure.

Contents

The PDB database is updated weekly (on Tuesday). Similarly, the PDB Holdings List is also updated weekly. As of 1 June 2010, the breakdown of current holdings was as follows:

Table 4.1

Experimental method	Proteins	Nucleic acids	Protein/Nucleic acid complexes	Other	Total
X-ray diffraction	53092	1210	2480	17	56799
NMR	7334	910	155	7	8406
Electron microscopy	200	17	76	0	293
Hybrid	18	1	1	1	21
Other	125	4	4	13	146
Total:	60769	2142	2716	38	**65665**

46,195 structures in the PDB have a structure factor file.

5,698 structures have an NMR restraint file.

These data explain that most structures are determined by X-ray diffraction, but about 15% of structures are now determined by protein NMR, and a few are even determined by cryo-electron microscopy.

SCOP

The Structural Classification of Proteins (SCOP) is a comprehensive database description of the structural and evolutionary relationships among all proteins whose 3-D structure is known. This database takes the information present in PDB and adds a layer of information in terms of analysis. SCOP contains domains present in these databases, which are hierarchically classified in to species, proteins, families, superfamilies, folds and classes. Families in SCOP are organized based on two factors: having residues that are 30% or greater identity and residues having lower identity but greater functional resemblance, *e.g.*, globulins of 15% sequence identity. A fold is defined in SCOP as a collection of superfamilies having the same secondary structures arranged in the same topological manner. Folds have been classified in to five classes: all alpha, all beta, alpha or beta, alpha and beta, and multidomain folds. A domain is considered to be an element of a protein which is self stabilizing and folds independently of the rest of the protein chain. Domains are not unique to proteins, nor to the products of a gene or gene family, thus are found distributed in a variety of proteins. These domains arise mainly due to the intrinsic physical and chemical properties of the proteins, which in turn form the bases of protein folding.

MMDB

The macromolecular 3-D structure database of NCBI's is called MMDB or Molecular Modelling Database. It is also known as the structural division of NCBI. It was designed to archive structural data from PDB as well as biomolecule generated by electron microscopy. MMDB is linked to the rest of the NCBI databases. An MMDB entry contains 3-D biopolymer or biomolecular structures including proteins and polynucleotides that are experimentally determined from X-ray crystallography and NMR spectroscopy. MMDB 3-D structure data is obtained from the PDB. The records of MMDB are created after sequence validation.

CATH

CATH is numerically indexed hierarchy of protein domain structures. CATH has four major classes: class(C), architecture (A), topology (T), and homologous (H), so the name CATH. It currently stores

34287 protein domain structures grouped into 1383 super families and 3285 sequence families. The basis of classification in CATH is the domain. Domains are arrived at using atleast three different methods. Proteins that have significant sequence similarity to a protein already in the databases inherit the domain boundaries. The first category "class" (C) is assigned automatically to more than 90% of the proteins data present in this database. The next category "architecture" (A) is assigned manually and gives information about the gross orientation of secondary structure. The next category "topology" (T) groups the structures according to their topological connection. Last category of this database, "homologous super family" (H), regroups proteins with highly similar structure and functions.

4.5 RETRIEVAL OF BIOLOGICAL DATA

Biological information is accessible in a huge amount over the World Wide Web (www), but the data are extensively distributed, and it is therefore required for scientists to have competent mechanisms for data retrieval. There are number of fanatical data retrieval tools are available. The most widely used of these are Entrez, DBGET and SRS. Each of these tools allows text based searching of a number of linked databases as well as sequence searching with utilities such as BLAST.

Entrez

Entrez is an integrated and World Wide Web based data retrieval system. It was developed by National Centre for Biotechnology Information (NCBI), a component of the United States National Library of Medicine, which integrates information, held in all NCBI databases. It is retrieval system of databases include nucleotide sequences, protein sequences, macromolecular structures, taxonomy and whole genomes, and biomedical literature via PubMed. Each category has one or more databases that support specific purposes. The sequence sources are diverse for the database sources, which include SWISS-PROT, Protein Identification Resources, PDB, GenBank protein translation, and RefSeq. Through PubMed one can access abstract, differences with links to the full text of the journals available on the web. There are embedded links leading to NCBI taxonomy. Further, Entrez provides wide links within and among database records. In their simplest form, these links may be simple cross-references among a sequence and the abstract of the paper in which it is reported, or between a protein sequence and its coding DNA sequence or perhaps its 3 D structure.

DBGET

It is an integrated data retrieval system developed and jointly maintained by the Institute for chemical Research (Kyoto University) and the Human Genome Centre (university of Tokyo). It is integrated with more than 20 databases, which can searched one at a time or in combination using the commands **bfind** (for text searches) or **bget** (for searches based on accession number). Hits are presented as a list of results together with any available associated information. **LinkDB** is an associated database of links (binary relationship) between entries in the different databases available to DBGET and further organism specific databases, such as the *Caenorhabditis Elegans* database (ACeDB) Flybase and the *Saccharomyces* Genome Database (SGD). DBGET is closely associated with KEGG, the Kyoto Encyclopedia of Genes and Genomes, which is maintained by the same group.

SRS (Sequence Retrieval System)

Sequence Retrieval System is a retrieval tool developed by the European Bioinformatics Institute (EBI). It is web base server system that allows the querying of biological sequence (DNA and protein sequence) database through a standardised interface. The user can use SRS to do both simple (single database) and more complex (multiple databases) queries. The main potency of SRS is its ability to allow user to plot a route between databases gathering related information from each one on a particular topic. SRS has several potential advantages:

- SRS is easier to use and more flexible than Entrez. For researchers, it is fast access to different life science data- genetic, protein, molecular, cellular and clinical.
- Integration of public and proprietary data through one interface.
- Rapid string search of large volume of data.
- Exclusive capability to perform cross-database queries.
- Many SRS servers provide access to the EMBLE database, which contains the same data as GenBank.

The SRS systems contain 175 data libraries. Those are divided into 16 library category:

1. Litreature,bibliography and reference database
2. Nucleotide sequence database
3. Protein structure database
4. Protein function database
5. Other protein sequence database
6. Deprecated protein databases
7. UniProt universal protein resource
8. Enzymes, reaction and metabolic pathway databases
9. Mutation and SNP databases
10. Mapping databases
11. Gene ontology resources
12. Other databases
13. EMBOSS result database
14. User owned databanks
15. Nucleotide related database
16. Application result database

4.6 SEQUENCE FILE FORMATS

All the programs employ the same format for storing sequences; it includes the database, query and numerous sequences. Many bioinformatic databases and software applications are designed to work with sequence data, and this requires a standard format for inputting nucleic acid and protein sequence information. Most popular and usable formats are: ALN, GenBank flat file, Plain/Raw, GDE, MSF, FASTA, NEXUS, PIR/NBRF, UniProt/ SwissProt, etc. Some are discussed below:

FASTA Format

FASTA is one of the simplest and well known sequence formats because it contains plain sequence information *i.e.*, readable by various bioinformatics analysis programmes. It has a single definition line that begins with a right angle bracket (>) followed by a sequence name. The rest of the header line is the description of the sequence, *i.e.*, version number, etc. The remaining lines contain the sequences (DNA/ Protein) in single letter codes. The blank space or gaps in a FASTA file are ignored. An optional (*) at the end of the file indicates end of the sequence. An example of the FASTA file containing a single sequence is as follows:

>Naveen 1

CCATCGTAGTAGCTGGTGCTAGGTGCCGTGGATAGTAGGTCAAGACGTAGATG
CATCGTGACTCGACGT

Multi FASTA format

FASTA file containing multiple sequences listed one after another are known as multi FASTA format. This format is accepted by various multiple sequence alignment programmes like Clustal W, Multalin etc. An example of a FASTA files containing multiple sequences:

>Naveen

GGCGGTTTTAAGCCGTAGTAGTCCGTGTCACGTCGGCTAGCTAGTACGTCACG
ACTAGTAGTCGTCGTC

>Shubha

AAACCGGAATTTTGTAGTACTATGCTAGTCAATGCCTGACCTGATCCGGACCGTAGTAT
GCGTGACGTGACGTGATG

>Shubha1

GCTTCTATCTAGGATCGATGCTGATCGATGCTGTAAGCTGATCGATCGTAGCT
AGCTGATCGTACGTCGATGCTGTT

PDB format

The PDB format is the extensively used molecular file format. This is the molecular file format of Protein Data Bank (PDB), maintained by Research Collaboratory for Structural Bioinformatics (RCSB). PDB formats contain lines of information in a text file. Every line of information in the file is called record. All the records are grouped into different sections (see Table 4.2). Most of the information in that database pertains to proteins, and the PDB format accordingly provides for rich description and annotation of protein properties. However, proteins are often crystallized in association with other molecules or ions such as water, ions, nucleic acids, drug molecules and so on, which therefore can be described in the PDB format as well.

Table 4.2 Different sections in PDB file format with their descriptions and records

Sections	Description	Records
Title	Summary descriptive remarks	HEADER, OBSLTE, TITLE, CAVEAT, COMPND, SOURCE, KEYWDS, EXPDTA, AUTHOR, REVDAT, SPRSDE, JRNL
Remarks	Bibliography, refinement, annotation	REMARKs 1, 2, 3, etc.

Primary structure	Peptide and nucleotide sequence and the relationship among the PDB sequence and that found in the sequence database(s)	DBREF, SEQAV, SEQRES, MODRES
Heterogen	Description of non-standard groups	HET, HETNAM, HETSYN, FORNUL
Secondary structure	Description of secondary structure	HELIX, SHEET, TURN
Connectivity annotation	Chemical connectivity	SSBOND, LINK, HYDBND, SLTBRG, CISPEP
Miscellaneous features	Features within the large biomolecules	SITE
Crystallographic	Description of the crystallographic cell	CRYST1
Coordinate transformation	Coordinate transformation operators	ORIGXn, SCALEn, MTRIXn, TVECT
Coordinate	Atomic coordinate data	MODEL, ATOM, SIGATM, ANISOU, SIGUIJ, TER, HETATM, ENDMDL
Connectivity	Chemical connectivity	CONECT
Book-keeping	Summary information, end of file marker	MASTER, END

Nexus file format

Nexus file format is extensively used in Bioinformatics. Several popular bio-software such as PAUP, GDA, MrBayes , Mesquite, and MacClade use this format. This file format begins with '#NEXUS' at the top of the file. Then, the file is separated in sections beginning with the word 'begin keyword' and ending the 'end'. The NeXus file format may also refer to a data format used in neutron, x-ray and muon scattering sciences.

Clustal format

Clustal is a widely used multiple sequence alignment computer program available either as a stand-alone or on-line programme. The stand-alone programme, which runs on UNIX and Macintosh, has two variants, ClustalW and ClustalX. The W version provides a simple text based interface and the X version provides a more user friendly graphical interface. It is accessible for Windows, Mac OS, and Unix/Linux. One of the most significant features of this programme is the flexibility of using substitution matrices. This program accepts an extensive range on input format incorporated NBRF/PIR, FASTA, EMBL/Swiss-Prot, Clustal, GCC/MSF, GCG9 RSF and GDE. The output format can be one or numerous of the ensuing: Clustal, NBRF/PIR, GCG/MSF, PHYLIP, GDE, and NEXUS.

GenBank flat-file format

GenBank, the genetic sequence database of National Centre for Biotechnology Information (NCBI), has its own sequence format known as the GenBank flat-file format. In flat-file every GenBank entry includes a brief description of the sequence, the scientific name and taxonomy of the source organism, and a table of features that identifies coding regions and other sites of biological importance, like transcription units, site of modification/ mutation, and repeats. GenBank flat-file is divided into three parts: Header, Features and Sequence. The introductory part of the flat-file is header, which gives detailed information about the source organism, length and type of the sequence, related literature etc. The features part is the annotated part of the sequence which lists position of the different genes present along with the related literature resource through Medline ID. The sequence part represent sequence composition denoted symbolically as A (Adenine), T (Thymine), G (Guanine), C (Cytosine).

EMBL Format

The EMBL format is the sequence format of European Molecular Biology Laboratory database. Every EMBL sequence file begins with an Identification Line (ID) and ends with a terminator line (//). EMBL file-format is composed of various kinds of lines, each with its own format. These are used to record the diverse types of data which make up the entry.

Plain Sequence Format

A sequence in plain format may contain only IUPAC characters and spaces (no numbers). A file in plain sequence format may only contain one sequence, while most others formats accept several sequences in one file.

RAW Format

This format is same as Plain format except it removes any white space or digits, accepts only alphabetic characters and rejects anything.

UniProt KB/Swiss-Prot format

UniProtKB/ Swiss-Prot, an annotated protein sequence database, follows a particular sequence format, which is termed UniProtKB/ Swiss-Prot format. This sequence format is almost similar to EMBL format.

PIR/NBRF format

PIR/NBRF is a National Biomedical Research Foundation format (Washington DC). The PIR format is analogous to FASTA format, however with significant differences. The first line of every sequence entry begins with a 'greater than' (>) sign, followed by a two letter code to indicate the type of sequences, then a semicolon, and a unique name of the entry in about 4–6 characters.

4.7 DATA ANALYSIS TOOLS

Data analysis tools available for nucleotide and peptide sequences.

4.7.1 Tools for Nucleotide Sequence Analysis

Various tools are available for nucleotide sequence analysis. Most of them are available freely on the web and many are downloadable also. There are many tools for nucleotide sequence analysis, some are given on next page:

Transeq: it is the tool for translating nucleic acid sequence into the corresponding peptide sequence.

CpGPlot/CpGReport

This tool belongs to the European Molecular Biology Open Software Suite (EMBOSS). CpGPlot is a tool used to plot the CpG rich areas of the nucleotide sequence. CpGReport produces an EMBL formatted report with a feature table that contains a key for rich island found, along with the location/qualifiers that depict the position of the island, its size, the total sum of C+G's in the island, the % CG, and the maximum observed value for it.

GCUA

Graphical Codon Usage Analyser is web-based tool that displays the codon prejudice in a graphical manner.

BLAST

Basic Local Alignment Search Tool is a tool for local alignment of sequences. Altschul, et al. 1990, created the BLAST group of programmes.

- BLAST ® (Basic Local Alignment Search Tool) is a set of similarity search programmes for protein/DNA, designed to explore all of the available sequence databases.
- BLAST algorithm seeks local as opposed to global alignments and is, therefore, able to detect relationships among sequences which share only isolated regions of similarity.
- BLAST is more than a tool to view sequences aligned with each other or to find homology, but a programs to locate regions of sequence similarity with a view to comparing structure and function.

Types of BLAST

The NCBI-BLAST programmes include the following:

- BLASTpgp
- MEGA BLAST
- RPS BLAST
- BLASTclust
- BLASTall
  - BLASTp
  - BLASTn
  - BLASTx
  - tBLASTn
  - tBLASTx
- formatdb
- fastacmd
- bl2seq

ORF Finder

The ORF (open reading frame) Finder is an essential analysis tool that finds all open reading frames of the query sequence. It uses the standard or alternative genetic codes to recognize all open reading frames. It is also useful in preparing a complete and accurate sequence during sequence submission.

Vecscreen

It is a tool developed by NCBI. It was developed with the intention to clean up public sequence databases and look for accidental inclusion of vectors in them. Some commonly used vectors are plasmids, phages, BAC, PAC, YAC, and cosmids.

4.7.2 Tools for Protein Sequence Analysis

There are various tools available for protein sequence analysis. Almost all of them are available freely on the web and many are downloadable too.

PSI-BLAST

Position-specific iterated BLAST (PSI-BLAST) is used for performing BLAST on the set of hits resulting from BLASTp to increase the sensitivity of the result. It computes a position-specific scoring matrix (PSSM*) that yields high scores to highly conserve positions and low scores to weakly conserve positions. Unlike other tools, PSI-BLAST runs as one programme starting with a single protein sequence and the intermediate steps of multiple alignments and profile construction invisible to the user.

** PSSM is a profile that is generated by calculating position-specific score for each position in the alignment.*

BLASTp

BLASTP is a tool for comparing an amino acid query sequence with a protein sequence database. It detects regions with similar sequences even in functionally unrelated proteins.

RPS BLAST

It searches a query sequence against a database of profiles, or score matrices, producing BLAST-like output.

tBLASTx

Basic Local Alignment Tool is a tool for local alignment of sequences. It compares the six-frame translations of nucleotide query sequence against six-frame translations of nucleotide sequence database using the BLAST algorithm.

CDART

CDART (conserved domain architecture retrieval tool) is mainly a comparative sequence analysis used to exhibit the functional domain that make-up a protein and lists proteins with analogous domain architectures when presented with a protein query sequence. It determines the domain architecture of a query protein sequence by comparing it to the database of conserved domain alignments. Since protein domain may be considered basic units of molecular function, proteins associated by domain architecture may play similar roles in cellular processes. Therefore, CDART serves as a valuable tool.

REVIEW QUESTIONS

1. Define bioinformatics and its applications.
2. What is NCBI? When it was established?
3. What are biological databases? Explain it.
4. Give the name of retrieval tools for biological databases.
5. What is sequence file format? Explain its different types.
6. Explain the Tools for protein sequence analysis.
7. Explain the Tools for nucleotide sequence analysis.

5

GENOMICS

Genomics is the application of the technology of high throughput molecular biology to many problems. It is the field of genetics that attempts to understand the content, organization, and function of genetic information contained in entire genomes. Genomics includes intensive efforts to determine the entire *DNA sequence* of organisms and fine-scale *genetic mapping* efforts. Genomics consists of two corresponding fields: structural genomics and functional genomics. The determination of organization and sequence of the genetic information contained within a genome is known as **Structural genomics**, and **functional genomics** characterizes the function of sequences explain by structural genomics. A third area, **comparative genomics,** compares the gene content, function, and organization of genomes of various organisms. The field of genomics is at the cutting edge of modern biology; information resulting from research in this field has made significant contributions to agriculture, human health, and various other areas. It has also provided gene sequences essential for producing medically important proteins through recombinant DNA technology. Comparisons of genome sequences from different organisms are leading to an improved perceptive of development and the history of life. Before embarking in detail study about genomics we should know about gene, and genome. A genome is the sum total of all an individual organism's genes. Thus, genomics is the study of all the genes of a cell, or tissue, at the DNA (genotype), mRNA, or protein (proteome) levels. In 1920, H. Winkler coined the term *genome to* describe 'the complete set of chromosomal and extra chromosomal genes of an organism or virus'. The term genomics is used primarily by Thomas Roderick in 1986, after a gap of 66 years, and he describes it as s scientific discipline of mapping, sequencing and analyzing the genome.

5.1 GENE AND GENOME

Gene

The gene is the functional unit of heredity, was defined by Mendel more than a century ago, who noticed that biological variations are inherited from parent organisms as specific, discrete traits. Every gene is a sequence within the genome that functions by giving rise to a discrete product (RNA or protein). Mendel summarized in his two laws, first the gene was recognized as a "particulate factor" that passes unchanged from parent to progeny. A gene may exist in alternative forms, these forms are called alleles. In diploid organisms, which have two sets of chromosomes, one copy of each chromosome is inherited from each parent. This is the same manner that is displayed by genes. One of the two copies of each gene is the paternal allele (inherited from the father); the other is the maternal allele (inherited from the mother).The similarity led to the discovery that chromosomes in fact carry the genes. Each chromosome consists of a linear set of genes. Each gene present at a particular position on the chromosome. The location is more formally called a genetic locus. The alleles of a gene are the different forms that are found at its locus. The organization of genes into chromosomes shown Fig. 5.1, was the innovation of genetic linkage; the tendency for genes on the same chromosome to remain together in the progeny instead of assorting independently as predicted by Mendel's laws. Once the unit of recombination (reassortment) was introduced as the measure of linkage, the creation of genetic maps became possible.

* More about DNA, consult chapter no.2

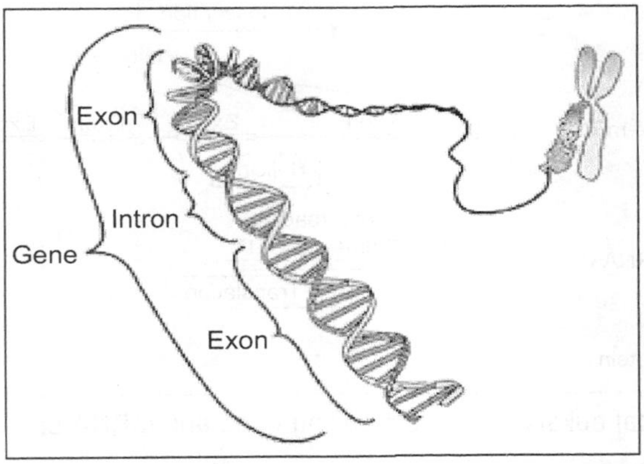

Fig. 5.1 A gene in relation to the double helix structure of DNA and to a chromosome (right)

Functional Structure of a Gene

Every gene includes regulatory regions, in addition to regions that clearly code for a RNA or protein. A regulatory region integrated by almost all genes is known as the promoter, which provides a site that is recognized by the transcription machinery when a gene is about to be transcribed and expressed. A gene can contain more than one promoter, ensuing in RNAs that differ in how far they extend in the 5' end. Though promoter regions contain a consensus sequence that is the generally common sequence at this position, few genes have "strong" promoters that bind the transcription machinery satisfactory, and others have "weak" promoters that bind weakly. These weak promoters generally allow a lower rate of transcription than the strong promoters, because the transcription machinery binds to them and initiates transcription less frequently. Other possible regulatory regions include enhancers, which can compensate for a weak promoter. The upstream regulatory regions are towards the 5′ end of the transcription initiation site. In contrary to prokaryotic promoter region, the eukaryotic promoter regions are very complicated and not easy to recognize. Operons have a various prokaryotic genes, or groups of genes whose products have related functions, which are transcribed as a unit. When the details of transcription were first carried out, scientists were staggered to find out that genes are interrupted by stretches of DNA that do not enclose protein coding information, called introns. Introns are interspersed between exons, protein coding sequences of a gene. Introns are common in eukaryotic genes but are rare in bacterial genes. Introns and exons are copied during transcription of mRNA, however, after transcription, the introns are removed by splicing and the exons are joined to yield the mature RNA (Fig. 5.2). There are two key steps involves in gene expression in to all organisms; the DNA containing gene must be transcribed from DNA to messenger RNA (mRNA) first; and, second, it must be translated from mRNA to protein. As shown in Fig. 5.2. The process by which a gene gives rise to a biologically functional molecule of either RNA or protein is called gene expression, and the resulting molecule itself is called a gene product.

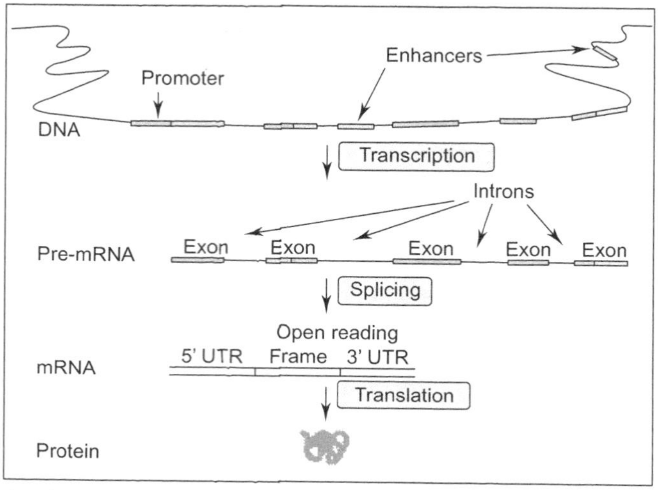

Fig. 5.2 Typical eukaryotic protein-coding gene and mRNA processing

Genes that encode proteins are composed of a series of three-nucleotide sequences called codons, which serve as the words in the genetic language. The relationship between a sequence of DNA and the sequence of the corresponding protein is called the genetic code. It is specifies the correspondence during protein translation between codons and amino acids. The genetic code is more or less the same for all known organisms. The structure and enzymatic activity of a protein follow from its primary sequence of amino acids. By determining the sequence of amino acids in a protein, the gene is able to carry all the information needed to specify an active polypeptide chain. In this way, a single type of structure; the gene, is capable to direct construction of immeasurable polypeptide forms.

The standard genetic code is shown in the following Tables 5.1; Shows what codons specify each of the 20 standard amino acids involved in translation. It is remarkable that the standard genetic code has basic features of error correction. Numerous codons which differ by only one base codes for the same amino acid, and most frequently the base that differs is the last base, which happens to be the base which is most often misread in the translation process. Furthermore, amino acids which tend to occur more frequently in proteins on average tend to have more codons which code for them.

Table 5.1 Codon Table This table shows the 64 codons and the amino acid each codon codes for.

		2nd base			
		U	C	A	G
1st base	U	**UUU** Phenylalanine **UUC** Phenylalanine **UUA** Leucine **UUG** Leucine, Start	**UCU** Serine **UCC** Serine **UCA** Serine **UCG** Serine	**UAU** Tyrosine **UAC** Tyrosine **UAA** Ochre (Stop) **UAG** Amber (Stop)	**UGU** Cysteine **UGC** Cysteine **UGA** Opal (Stop) **UGG** Tryptophan
	C	**CUU** Leucine **CUC** Leucine **CUA** Leucine **CUG** Leucine, Start	**CCU** Proline **CCC** Proline **CCA** Proline **CCG** Proline	**CAU** Histidine **CAC** Histidine **CAA** Glutamine **CAG** Glutamine	**CGU** Arginine **CGC** Arginine **CGA** Arginine **CGG** Arginine
	A	**AUU** Isoleucine, Start[2] **AUC** Isoleucine **AUA** Isoleucine **AUG** Methionine, Start[1]	**ACU** Threonine **ACC** Threonine **ACA** Threonine **ACG** Threonine	**AAU** Asparagine **AAC** Asparagine **AAA** Lysine **AAG** Lysine	**AGU** Serine **AGC** Serine **AGA** Arginine **AGG** Arginine
	G	**GUU** Valine **GUC** Valine **GUA** Valine **GUG** Valine, Start[2]	**GCU** Alanine **GCC** Alanine **GCA** Alanine **GCG** Alanine	**GAU** Aspartic acid **GAC** Aspartic acid **GAA** Glutamic acid **GAG** Glutamic acid	**GGU** Glycine **GGC** Glycine **GGA** Glycine **GGG** Glycine

1. The codon AUG both codes for methionine and serves as an initiation site.
2. This is a start codon for *prokaryotes* only.

Genome

An entire set of genetic information for any organism is its genome, and every genomes are encoded in nucleic acids; either DNA or RNA. The coding system for genomic information also is common to all life. The term was adapted by Hans Winkler, Professor of Botany at the University of Hamburg, Germany in 1920. the enormous majority of genes are positioned on a single chromosome of circular DNA in prokaryotes, whereas eukaryotes generally possess multiple individual linear DNA helices packed into dense DNA-protein complexes called chromosomes. Cells or organisms with only one copy of every chromosome are called haploid; and with two copies of chromosome are called diploid. Cells or organisms with more than two copies of chromosomes are called polyploid.

The word genome is applied exclusively to indicate that stored on a complete set of nuclear DNA (*i.e.*, the "nuclear genome") but it can also be applied to that stored within organelles such as mitochondria, chloroplast and some other cell organelles. Mitochondria that contain their own DNA, which codes for some of the proteins and RNA molecules found in the organelles. Mitochondrial DNA (mtDNA) has a lot of advantages for the study of evolutionary relationships.

1. The Mitochondrial DNA (mtDNA) has short length of the DNA molecule than the length of DNA in the chromosomes found in the nuclei of eukaryotic cells.

2. Mitochondrial DNA (mtDNA) is copious because each cell typically has numerous mitochondria, each with several copies of the mitochondrial chromosome. Therefore, mtDNA is easier to segregate and study than is nuclear DNA.
3. Mitochondrial DNA (mtDNA) in animals tends to develop more rapidly than nuclear DNA, and so it's useful for looking at relationships among closely related organisms.

Finally, mtDNA is typically inherited from single parent (generally the mother), and so its genes are not reshuffled every generation by recombination, which tends to obscure genetic relationships.

Geneticists have long documented that numerous traits associated with chloroplasts reveal cytoplasmic inheritance, indicating that these traits are not encoded by nuclear genes. Chloroplasts were exposed to have their own DNA In 1963. The chloroplast genome ranges in size from 80,000 to 600,000 bp, among different plants, but most chloroplast genomes range from 120,000 to 160,000 bp. Chloroplast DNA is generally contained on a single, double stranded DNA molecule that is circular, is highly coiled, and lacks associated histone proteins. As in mtDNA, multiple copies of the chloroplast genome are found in each chloroplast, and there are multiple organelles per cell; so there are several hundred to several thousand copies of cpDNA in a typical plant cell.

The number of genes in the genome can be identified directly by defining open reading frames. Extensive mapping of this nature is difficult by the fact that interrupted genes may consist of various separated open reading frames. Because we do not necessarily have information about the functions of the protein products, or indeed proof that the open reading frames are expressed at all, this approach is restricted to defining the potential of the genome.

An additional approach is to define the lot of genes directly in terms of the transcriptone (by directly identifying all the mRNAs) or proteome (by directly identifying all the protein). This reveals how many genes are expressed in a specific tissue or cell type, what difference exists in the relative levels of expression, and how many of the genes expressed in one particular cell type are unique to that cell type and how many are also expressed elsewhere.

Eukaryotic genome contains both nonrepetitive and repetitive DNA sequences. Repetitive DNA sequence that is present in more than one copy in each genome. It is having moderately and highly repetitive sequences. The general character of the eukaryotic genome can be assessed by the kinetics of reassociation of denatured DNA. Eukaryotic genome can be divided in to three types of DNA sequences:

- Nonrepetitive sequences are unique; the haploid genome contains only one copy of such a sequence.
- Moderately repetitive sequences are found in multiple copies. The copies may be corresponding to one another or, more typically, may be seen to be related although not identical.
- Highly repetitive sequences are very short and are present in large numbers of copies, often organized as tandem repeats. These include very large blocks of material organized as satellite DNA and smaller blocks that are called minisatellites or microsatellites, depending on their length.

5.2 HISTORY OF GENOMICS

Genomics was recognized by *Fred Sanger* when the genome of mitochondria and virus was completely sequenced by him. The techniques of sequencing, genome mapping, data storage, and bioinformatic analyses are recognized by his group in the 1970–1980s. The genomics is still concerned with *sequencing* the genomes of various organisms, but the knowledge of full genomes has created the possibility for the field of *functional genomics,* primarily concerned with patterns of *gene expression* during various conditions. The most significant tools are *microarrays* and *bioinformatics* for genetic analysis of genome. A new scientific discipline, **proteomics**, is committed to understanding the complex relationship of disease and protein expression. In proteomics, the ***proteome*** is the entire set of proteins. It should correspond to the mRNAs in the transcriptome, though there can be differences of detail reflecting changes in the relative abundance or stabilities of mRNAs and proteins. It can be used to refer to the set of proteins coded by the whole genome or produced in any particular cell or tissue. Dr. Tom Roderick, a geneticist at the Jackson Laboratory, have been coined the actual term genomics; over beer at a meeting held in Maryland on the mapping of the human genome in 1986. In 1972, *Walter Fiers* and his team at the Laboratory of Molecular Biology of the *University of Ghent* (*Ghent, Belgium*) were the first to determine the sequence of a gene: the gene for *Bacteriophage MS2* coat protein. In 1976, the team determined the complete nucleotide-sequence of bacteriophage MS2-RNA. In 1977, *Frederick Sanger* was sequenced first DNA-based genome of *bacteriophage X174;* (5,368 *bp*). As of September 2007, the complete sequence was known of about 1879 *viruses, 577 bacterial* species and roughly 23 *eukaryote* organisms, of which about half are *fungi*. Most of the bacteria whose genomes have been completely sequenced are problematic disease-causing agents, such as *Haemophilus influenzae*. Another sequenced species, most were selected because of well-studied model organisms or promised to become good models. Yeast (*Saccharomyces cerevisiae*) has long been an important *model organism* for the *eukaryotic cell,* while the fruit fly *Drosophila melanogaster* has been a very important tool (notably in early pre-molecular *genetics*). The worm *Caenorhabditis elegans* is an often used simple model for *multicellular organisms.* The zebrafish (*Brachydanio rerio)* is used for many developmental studies on the molecular level and the flower *Arabidopsis thaliana* is a model organism for flowering plants. The *Japanese pufferfish* (*Takifugu rubripes*) and the *spotted green pufferfish* (*Tetraodon nigroviridis*) are interesting because of their small and compact genomes, containing very little non-coding DNA compared to most species. The mammals dog (*Canis familiaris*), brown rat (*Rattus norvegicus*), mouse (*Mus musculus*), and chimpanzee (*Pan troglodytes*) are all significant model animals in medical research.

5.3 INTRODUCTION TO GENOME SEQUENCING PROJECTS

During the mid 1980s, the U.S. department of energy started several projects to develop the human genome. They got success when, in 1995, the complete genomes of two microorganism (two bacteria, *Haemophilus influenzae* and *Mycoplasma genitalicum*) were sequenced. In 1996, the first yeast *Saccharomyces cerevisiae* genome sequencing was completed. In 1997, sequencing of genomes of the two best studied bacteria, *Escherichia coli* and *Bacillus subtitles'* was completed. The marvellous discovery had been done in the field of genomics, when the working draft of the human genome was released in 2000 and it completely sequenced in 2003.

Introduction to Human Genome Projects

Methods for mapping and sequencing DNA fragments had been satisfactorily developed till 1980, that geneticists began sincerely proposing that the whole human genome could be sequenced. An international association was anticipated to undertake the Human Genome Project (Fig. 5.3); initial estimates suggested that 15 years and $3 billion would be required to accomplish the task. The project incorporated obtaining the sequence, identifying the genes, and describing their function. As a part of the effort, the genomes of several model organisms, including *Escherichia coli* (bacterium), *Saccharomyces cerevisiae* (yeast), *Drosophila melanogaster* (fruit fly), *Arabidopsis thaliana* (a plant), and *Caenorhabditis elegans* (a nematode) were to be sequenced as well. The genomes of these model organisms were sequenced to help develop methods that could then be applied to the sequencing of the human genome and to provide sequenced genomes with which to compare the organization and structure of the human genome. In October 1990, the Human Genome Project officially got underway. Initial efforts focused on developing new and automated methods for cloning and sequencing DNA and on generating comprehensive physical and genetic maps of the human genome. The methods described earlier for mapping, sequencing, and assembling DNA fragments were crucial in these early stages of the project. Large scale physical maps were completed by 1993 for all 23 pairs of human chromosomes. At the same time, automated sequencing techniques had been developed that made large-scale sequencing possible. The primary attempt to sequence the genome was a public project consisting of the international collaboration of 20 research groups and hundreds of individual researchers who formed the International Human Genome Sequencing Consortium. In 1998, Craig Venter announced that he would lead a company called Celera Genomics in a private effort to sequence the human genome. The public and private efforts moved ahead simultaneously but used different approaches.

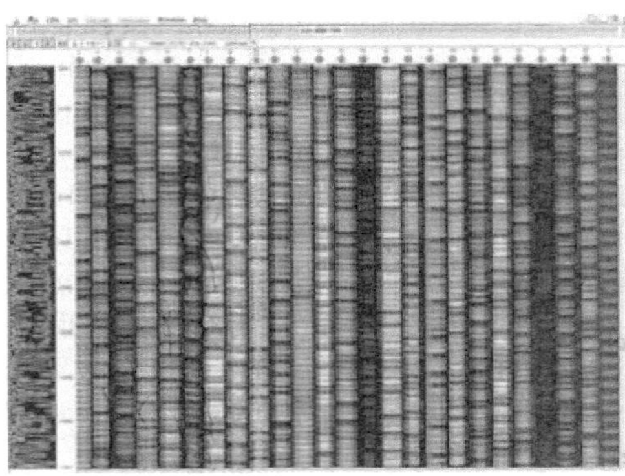

Figure 5.3 Methods for mapping and sequencing DNA fragments had been satisfactorily developed till 1980, that geneticists began sincerely proposing that the whole human genome could be sequenced. An international association was anticipated to undertake the Human Genome Project (Fig. 5.3); initial estimates suggested that 15 years and $3 billion would be required to accomplish the task. The project incorporated obtaining the sequence, identifying the genes, and describing their function. As a part of the effort, the genomes of several model organisms, including Escherichia coli (bacterium), Saccharomyces cerevisiae (yeast), Drosophila melanogaster (fruit fly), Arabidopsis thaliana (a plant), and Caenorhabditis elegans (a nematode) were to be sequenced as well. The genomes of these model organisms were sequenced to help develop methods that could then be applied to the sequencing of the human genome and to provide sequenced genomes with which to compare the organization and structure of the human genome. In October 1990, the Human Genome Project officially got underway. Initial efforts focused on developing new and automated methods for cloning and sequencing DNA and on generating comprehensive physical and genetic maps of the human genome. The methods described earlier for mapping, sequencing, and assembling DNA fragments were crucial in these early stages of the project. Large scale physical maps were completed

by 1993 for all 23 pairs of human chromosomes. At the same time, automated sequencing techniques had been developed that made large-scale sequencing possible. The primary attempt to sequence the genome was a public project consisting of the international collaboration of 20 research groups and hundreds of individual researchers who formed the International Human Genome Sequencing Consortium. In 1998, Craig Venter announced that he would lead a company called Celera Genomics in a private effort to sequence the human genome. The public and private efforts moved ahead simultaneously but used different approaches.

The Human Genome Sequencing Consortium used a map-based approach; many copies of the human genome were cut up into fragments of about 150,000 bp each, which were inserted into bacterial artificial chromosomes. Yeast artificial chromosomes and cosmids had been used in early stages of the project but did not prove to be as stable as the BAC clones, though YAC clones were instrumental in putting together some of the larger contigs (continuous sequences). Restriction fingerprints were used to assemble the BAC clones into contigs, which were positioned on the chromosomes with the use of genetic markers and probes. The individual BAC clones were sheared into smaller overlapping fragments and sequenced, and the whole genome was assembled by putting together the sequence of the BAC clones. Celera Genomics used a whole-genome shotgun approach to determine the human genome sequence, although the genetic and physical maps produced by the public effort helped Celera assemble the final sequence. In this approach, small-insert clones were prepared directly from genomic DNA and then sequenced. The overlapping of DNA sequences among these small-insert clones was then used to assemble the entire genome. Both public and private sequencing projects announced the completion of a rough draft that included most of the sequence of the human genome in the summer of 2000, five years ahead of schedule. Analysis of this sequence was published 6 months later. The human genome sequence was declared completed in the spring of 2003, although some gaps still remain. For most chromosomes, the finished sequence is 99.999% accurate, with less than one base-pair error per 100,000 bp, which is 10 times as accurate as the initial goal. The availability of the complete sequence of the human genome is proving to be of enormous benefit. It has made it easier to identify and isolate genes that contribute to lots of human diseases and to make probes that can be used in genetic testing, diagnosis, and drug development. The sequence is also providing significant information about many fundamental cellular processes. Comparisons of the human genome with those of other organisms are adding to our understanding of evolution and the history of life.

The "genome" of any given entity (except for *identical twins* and *cloned* organisms) is unique mapping "the human genome" involves sequencing multiple variations of each gene. The Human Genome (haploid) contains 22 autosomes plus the X or Y. The chromosomes range in size from 45–279 Mb of DNA, making a total genome content 3,286 Mb. On the basis of chromosomes structure, the whole genome is divided into two regions that are euchromatin and heterochromatin. The euchromatin comprises the majority of the genome, 4 2.9×10^9 bp. Scientists have learned remarkable facts from the human genome project:

- The human genome consists about 3.1 billion base pair; in which 2.85 base pairs have been fully sequenced.
- The genome is approximately 99.9% the same between individuals of all nationalities and back grounds.
- Less than 2% of the genome codes for genes.
- The human genome has approximately 20,000–25,000 protein coding genes.

- Repetitive DNA sequences account for at least 50% of the non coding DNA; it shows the immense majority of our DNA is non protein coding.
- ~ 60% of human genes are alternatively spliced.
- Functions for more than half of the entire human genes are unknown.
- Most of the human genome shows a high degree of sequence resemblance to genes in other organisms.
- Thousands of human disease genes have been recognized and mapped to their chromosomal locations.

Here the question arise that how genes and other sequences are distributed in the human genome?

The answer is, some chromosomes are very deprived in genes and have >25% of their sequences as 'desert'-regions longer than 500Kb where there are no genes. Most of the human genome is repetitive sequences. Repetitive sequences account for >50% of the human genome, as shown in Fig. 5.3, and divided into five regions.

- Transposons* (either active or inactive) account for vast majority (45% of the genome).
- Processed pseudogens (~3000 in all, account for ~0.1% of total DNA)
- Simple sequence repeats (highly repetitive DNA account for ~3%).
- Segmental duplications (account ~5%)
- Tandem repeats form blocks of one type of sequence.

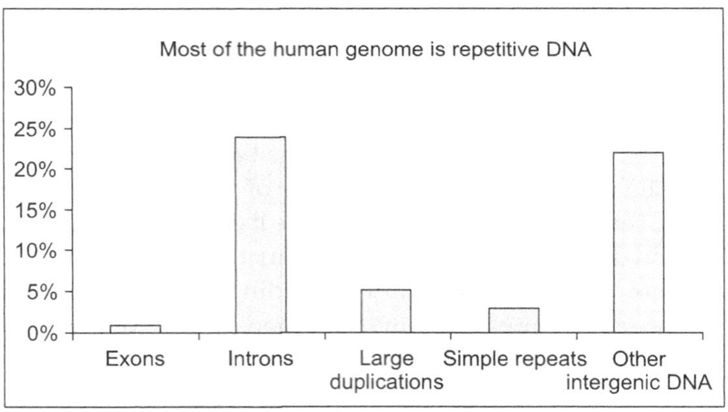

Fig. 5.3 The biggest component of the human genome consists of transposons. Other sequences include large duplication and simple repeats

* The sequence of the human genome emphasizes the importance of transposons. Transposons have the capacity to replicate themselves and insert into location. They may function exclusively as DNA elements.

5.4 GENE PREDICTION AND COUNTING

The purpose of gene prediction is to identify regions of genomic DNA that encode proteins. In the commencement of 1980, *in silico* gene prediction has evolved from simple methods based on coding regions statistics to classy methodologies. *In silico* development of gene predictions because of human genome project. The gene prediction assignment, start with a concise description of the biological experience of protein synthesis. In an organism; every chromosome contains a double-helix of nucleotide bases called deoxyribonucleic acid (DNA). Each nucleotide contains one of four bases: adenine (a), cytosine (c), guanine (g), and thymine (t). during the transcription process, the DNA helix is mRNA contains superfluous sequences called introns that are not needed in unwound and a contiguous subsequence of bases is copied to a new constituent called pre mRNA. These introns are detached in the splicing stage, and the remaining bases (exons) are reassembled into mRNA. The mRNA contains an adjacent subsequence of bases called the coding sequence (CDS) that are deterministically translated into amino acids. Translation maps base triplets (codons) into amino acids, which combine to form a protein. In gene prediction, we must predict the coding sequence from the original strand of DNA. Though, because not all of these bases contribute to protein manufacturing, it is difficult to determine which proteins will be generated from a random DNA sequence. We can use sequential structure to determine which proteins, if any, a subsequence of bases will encode. We refer to these protein coding regions as genes; consequently the task of gene prediction is to assume which subsequence of DNA correspond to genes. A popular probabilistic sequence model is the Hidden Markov Model (HMM), one such gene predicting tool is Genscan; which has widespread use in natural language processing tasks such as speech recognition, part-of-speech tagging, and information extraction. Noting the surprising similarities between DNA sequences and human language, bioinformatics researchers have effectively adapted HMMs for gene prediction. HMMs are prominent models because it is relatively straightforward for a machine to learn their parameters and for a human to interpret them. However, one drawback of HMMs is that it is cumbersome to model arbitrary, dependent features of the input sequence. This drawback of HMMs can be particularly troublesome for gene prediction, since there are many potentially valuable external sources of evidence (*e.g.*, genomic databases, EST data, and tiling arrays). HMMs assume that each feature is generated independently by some hidden process; however, this is in general not the case. The technique to tackle this problem is to explicitly model these dependencies by complicating the HMM structure, but this grows intractable as the number of features increases. To overcome this, there is the use of another sequence model that has become popular in language processing and vision tasks, the Conditional Random Field (CRF). CRFs have out-performed HMMs on language processing tasks such as information extraction and shallow parsing, and it show that similar performance gains can be obtained for gene prediction. Gene prediction requires a combination of algorithms with different types of biological databases. There are basically several algorithms for Gene prediction. They are given below:

- Homology based Gene prediction
- Gene prediction through detection of homology to know the proteins
- Gene prediction through comparison with Expressed Sequence Tags(EST) database
- Ab Initio Gene prediction
- Systemic Gene prediction.

5.5 GENOME SIMILARITY

When the work on Human genome project was in progress, at that time two questions were raised:

- Genome of individuals to be sequenced
- Percentage of genome similarity between two individuals.

Though all the humans are similar, human genome is about 99.8% similar to each other. The percentage of similarity refers to 'consensus human genome'. It means that 0.2% difference in genome makes them different from each other. There lies a difference of one in thousand nucleotides in the genomes of two different individual's .Only this percentage of DNA brings a uniqueness in them.

5.6 SNP'S (SINGLE-NUCLEOTIDE POLYMORPHISM)

Single-nucleotide polymorphism or SNP (pronounced 'snip') is a genetic variation among individuals, limited to a single base pair which can be substituted, inserted or deleted. Sickle-cell anaemia is an example of a disease caused by a specific SNP: an A '! T mutation in the â-globin gene changes a Glu'! Val, creating a sticky surface on the haemoglobin molecule that leads to polymerization of the deoxy form. SNPs are dispersed throughout the genome, occurring on the average every 2000 base pairs. Though they arose by mutation, numerous positions containing SNPs have low mutation rates, and provide stable markers for mapping genes. The defined set of SNPs and other genetic variants observed on a single chromosome or part of a chromosome is called a haplotype (Fig. 5.4). SNPs inside a haplotype are physically linked and then tend to be inherited together. When a SNP is physically close to a disease causing locus, it will tend to be inherited along with the disease-causing allele. Therefore the SNP marks the position of a genetic locus that causes the disease. A SNP can also be useful for determining family relationships. Since most SNPs are unique within a population, having arisen only once by mutation, the presence of the same SNP in two persons frequently indicates that they have a common ancestor. An international attempt, called the HapMap project, is currently underway to catalog SNPs and other genetic variants and to identify common haplotypes in human populations for use in linkage and family studies. The project will collect and identify 600,000 SNPs from several hundred DNA samples collected from four major human ethnic groups. Although there are maybe from 8 million to 10 million SNPs in the human population, they are present in a much smaller set of haplotypes, and therefore a much smaller number of SNPs perhaps only 100,000 can be used to identify all of the haplotypes.

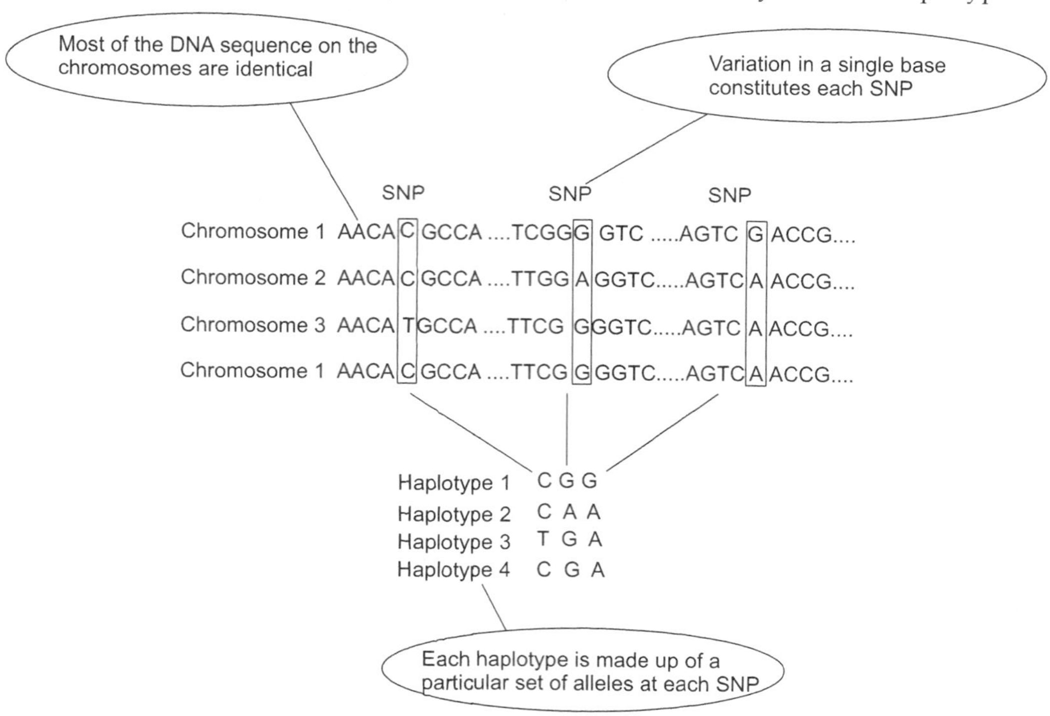

Fig. 5.4 The observation of single chromosome or part of a chromosome for haplotype

5.6.1 Types of SNPS

Types of SNPs

- Non-coding region
- Coding region
    - (*a*) Synonymous
    - (*b*) Nonsynonymous
        - Missense
        - Nonsense

Single nucleotides may be changed (substitution), removed (deletions) or added (insertion) to a polynucleotide sequence. Insertion or deletion SNPs may shift translational frame. Single nucleotide polymorphisms may fall within coding sequences of genes, non-coding regions of genes, or in the intergenic regions between genes. SNPs within a coding sequence will not necessarily change the amino acid sequence of the protein that is produced, due to degeneracy of the genetic code. A SNP in which both forms lead to the same polypeptide sequence is termed *synonymous* (sometimes called a silent mutation) — if a different polypeptide sequence is produced they are *nonsynonymous*. A nonsynonymous change may either be missense or nonsense, where a missense change results in a different amino acid, while a nonsense change results in a premature stop codon. SNPs that are not in protein-coding regions may still have consequences for gene splicing, transcription factor binding, or the sequence of non-coding RNA.

5.7 TYPES OF GENOMICS

The genomics is broadly categorized into following types:

- Structural Genomics
- Functional Genomics
- Comparative Genomics
- Nutritional Genomics

5.7.1 Structural Genomics

The structural genomics is concerned with DNA sequencing, sequence assembly, sequence organisation and management. Commonly, one of the early steps in characterizing a genome is to prepare genetic and physical maps of its chromosomes. These maps provide information about the relative locations and distance between genes, molecular markers, and chromosome segments, which are often essential for positioning chromosome segments and aligning stretches of sequenced DNA into an entire genome sequence. Genetic maps are based on rates of recombination and are measured in percent recombination, or centimorgans. Physical maps are based on the physical distances and are measured in base pairs.

5.7.2 Functional Genomics

Independently use of genomic sequence is limited. Thus the functional genomics provide the novel information about the genome. Functional genomics is, in essence, probing genome sequences for implication; identifying genes, recognizing their organization, and understanding their function.

The objects of functional genomics comprise the identification of all the RNA molecules transcribed from a genome, called the transcriptome of that genome, and all the proteins encoded by the genome, called the proteome. Functional genomics exploits both bioinformatics and laboratory based experimental approaches in its search to define the function of DNA sequences.

5.7.3 Comparative Genomics

Comparative genomics compares the content and organization of entire genomic sequences from different organisms. The purpose is to gain a better understanding of how species have evolved and to determine the function of genes and noncoding regions of the genome. Genome researchers look at various different characters when comparing genomes: sequence similarity, gene location, the length and number of coding regions (called exons) within genes, the amount of noncoding DNA in each genome, and highly conserved regions maintained in organisms as simple as bacteria and as complex as humans. Prokaryotic genomes are small, usually ranging from 1 million to 3 million base pairs of DNA, with several thousand genes. Among multicellular eukaryotic organisms, there is no clear relation between organismal complexity and amount of DNA or gene number. A substantial part of the genome in eukaryotic organisms consists of repetitive DNA, much of which is derived from transposable elements. Many eukaryotic genomes have homologous genes in common. Comparative genomics involves the use of computer programs that can line up multiple genomes and look for regions of similarity among them. Some of these sequence-similarity tools are accessible to the public through the Internet. One of the most extensively used is BLAST, which is available from the National Centre for Biotechnology Information. BLAST is a set of programs designed to perform similarity searches on all available sequence data.

5.7.4 Nutritional Genomics

Nutritional genomics is a fast promising area of the genomics; in which we study about the manipulation of genes, which are responsible for the synthesis of nutritionally important enzymes or other molecules, often involving whole biosynthetic pathways. This will cover the way for inserting these genes in to crop plants to enrich them in special ways. Golden rice is the first plant treated as the example of nutritional genomics; in which the biosynthetic machinery for ?-carotene (pro-vitamin A) is introduced in to the rice genome (Oryza sativa) to express a new feature in the rice grain.

REVIEW QUESTIONS

1. What do you mean by a gene? Give the functional structure of gene.
2. What do you understand by genetic code? Explain it.
3. What do you understand by the term genome?
4. Give the brief history of genomics.
5. What is genomics? Give a brief account of genome sequencing project.
6. What do you understand by gene prediction and gene counting?
7. Write a note on genome similarity.
8. Explain the different type of genomics.
9. Write in brief about human genome project.
10. Write short note on SNP's.

INDEX

S

T

U

V

W

X

Z